Lecture Notes
in Business Information Processing   428

T0202509

More information about this series at http://www.springer.com/series/7911

José González Enríquez · Søren Debois ·
Peter Fettke · Pierluigi Plebani ·
Inge van de Weerd · Ingo Weber (Eds.)

Business Process Management

Blockchain and Robotic Process Automation Forum

BPM 2021 Blockchain and RPA Forum
Rome, Italy, September 6–10, 2021
Proceedings

 Springer

Editors
José González Enríquez (iD)
Escuela Técnica Superior de Ingeniería
Seville, Spain

Peter Fettke (iD)
German Research Center for Artificial
Intelligence (DFKI) and Saarland University
Saarbrücken, Germany

Inge van de Weerd (iD)
Utrecht University
Utrecht, The Netherlands

Søren Debois (iD)
IT University of Copenhagen
Copenhagen, Denmark

Pierluigi Plebani (iD)
Politecnico di Milano
Milan, Italy

Ingo Weber (iD)
TU Berlin
Berlin, Germany

ISSN 1865-1348   ISSN 1865-1356 (electronic)
Lecture Notes in Business Information Processing
ISBN 978-3-030-85866-7   ISBN 978-3-030-85867-4 (eBook)
https://doi.org/10.1007/978-3-030-85867-4

This Springer imprint is published by the registered company Springer Nature Switzerland AG
The registered company address is: Gewerbestrasse 11, 6330 Cham, Switzerland

Preface

This volume contains the proceedings of the Robotic Process Automation (RPA) Forum and the Blockchain Forum, which took place during September 6–10, 2021. Both of the forums were organized as part of the 19th International Conference on Business Process Management (BPM 2021), held in Rome, Italy.

The concept of Robotic Process Automation (RPA) has gained relevant attention in both industry and academia. RPA raises a way of automating mundane and repetitive human tasks requiring a lower level of intrusiveness with the IT infrastructure. The RPA Forum aimed to bring together researchers from various communities and disciplines to discuss challenges, opportunities, and new ideas related to RPA and its application to business processes in private and public sectors. The forum solicited contributions related to three main RPA areas: management, technology, and applications. The keynote given by Andrés Jiménez Ramírez from the University of Seville, Spain, revolved around two main topics, namely, how to frame RPA in the existing Business Process Management (BPM) paradigm and hyperautomation. The RPA Forum took place for the second time, after its first successful appearance at BPM 2020, in Seville, Spain.

A blockchain is a distributed data structure that guarantees immutability and integrity protection, providing a practical solution to complex problems in coordination. Blockchain-based systems open up diverse opportunities in the context of the BPM lifecycle to redesign business activities in a wide range of fields, including healthcare, supply chain, logistics, and finance. However, these opportunities come with challenges to security and privacy, scalability, and performance. The third Blockchain Forum provided a platform for the discussion of ongoing research and success stories on the use of blockchain, including techniques for and applications of blockchain and Distributed Ledger Technology. The program was complemented with a keynote by Fransisco Curbera from the IBM Center for Blockchain and Data Transparency.

The RPA Forum received five submissions, which led to the acceptance of the top three as full papers; the Blockchain Forum received nine papers, of which the top five were accepted as full papers. The overall acceptance rate was 57%. For both forums, each submission was reviewed by at least three members of the respective Program Committees.

We hope that the reader of these proceedings will enjoy the papers presented at both forums. We would like to congratulate both the authors of the accepted papers and those who submitted their work that, unfortunately, was not accepted despite its quality. We also thank our colleagues who acted as reviewers in the selection process and provided the authors with meaningful and constructive comments.

Finally, special thanks go to Massimo Mecella (general chair of BPM 2021) and his Organizing Committee, including Simone Agostinelli, Dario Benvenuti, Eleonora Bernasconi, Francesca de Luzi, Lauren Stacey Ferro, Francesco Leotta, Andrea Marrella,

Francesco Sapio, and Silvestro Veneruso. Their great effort enabled these forums take place under safe conditions.

July 2021

José González Enríquez
Søren Debois
Peter Fettke
Pierluigi Plebani
Inge van de Weerd
Ingo Weber

Organization

The 19th International Conference on Business Process Management (BPM 2021) was hosted by the Dipartimento di Ingegneria Informatica, Automatica e Gestionale Antonio Ruberti at Sapienza Università di Roma. It took place in Rome, Italy. The Robotic Process Automation Forum and Blockchain Forum were co-located with the main conference, which took place during September 6–10, 2021.

Executive Chair

General Chair

Massimo Mecella   Sapienza University of Rome, Italy

Robotic Process Automation Forum

Program Committee Chairs

Peter Fettke   German Research Center for Artificial Intelligence
                                (DFKI) and Saarland University, Germany
José González Enríquez   University of Seville, Spain
Inge van de Weerd   Utrecht University, The Netherlands

Program Committee

Sorin Anagnoste   Bucharest University of Economic Studies, Romania
Aleksandre Asatiani   University of Gothenburg, Sweden
Hyerim Bae   Pusan National University, South Korea
Christian Czarnecki   Hamm-Lippstadt University of Applied Sciences,
                                Germany
Carmelo del Valle Sevillano   University of Seville, Spain
Francisco J. Domínguez Mayo   University of Seville, Spain
Amador Durán Toro   University of Seville, Spain
María J. Escalona Cuaresma   University of Seville, Spain
Michael Fellmann   University of Rostock, Germany
Lukas-Valentin Herm   University of Wuerzburg, Germany
Florian Imgrund   digital& GmbH, Germany
Hannu Jaakkola   Tampere University, Finland
Christian Janiesch   Technical University of Dresden, Germany
Andrés Jiménez Ramírez   University of Seville, Spain
Jari Jussila   Häme University of Applied Sciences, Finland
Mathias Kirchmer   BPM-D, UK

Agnes Koschmider	Kiel University, Germany
Volodymyr Leno	The University of Melbourne, Australia
Esko Penttinen	Aalto University, Finland
Artem Polyvyanyy	The University of Melbourne, Australia
Milla Ratia	Tampere University, Finland
Hajo A. Reijers	Utrecht University, The Netherlands
Minseok Song	Pohang University of Science and Technology, South Korea
Rehan Syed	Queensland University of Technology, Australia
Jonas Wanner	University of Wuerzburg, Germany
Judith Wewerka	BMW, Germany

Blockchain Forum

Program Committee Chairs

Søren Debois	IT University of Copenhagen, Denmark
Pierluigi Plebani	Politecnico di Milano, Italy
Ingo Weber	TU Berlin, Germany

Program Committee

Marco Comuzzi	Ulsan National Institute of Science and Technology, South Korea
Claudio Di Ciccio	Sapienza University of Rome, Italy
Alevtina Dubovitskaya	Lucerne University of Applied Sciences and Arts (HSLU), Switzerland
José María García	University of Seville, Spain
Luciano García Bañuelos	Tecnológico de Monterrey, Mexico
Inma Hernández	University of Seville, Spain
Marko Hölbl	University of Maribor, Slovenia
Sabrina Kirrane	Vienna University of Economics and Business, Austria
Julius Köpke	Alpen-Adria-Universität Klagenfurt, Austria
Agnes Koschmider	Kiel University, Germany
Qinghua Lu	CSIRO, Australia
Raimundas Matulevicius	University of Tartu, Estonia
Jan Mendling	Wirtschaftsuniversität Wien, Austria
Giovanni Meroni	Politecnico di Milano, Italy
Alex Norta	Tallinn University of Technology, Estonia
Juan Pavón	Universidad Complutense de Madrid, Spain
Tijs Slaats	University of Copenhagen, Denmark
Mark Staples	CSIRO, Australia
Burkhard Stiller	University of Zurich, Switzerland
Ali Sunyaev	University of Cologne, Germany
Horst Treiblmaier	Modul University Vienna, Austria
Kaiwen Zhang	École de technologie supérieure, Canada

Contents

Robotic Process Automation Forum

Humans, Processes and Robots:
A Journey to Hyperautomation

Andrés Jiménez-Ramírez(✉) (iD)

Departamento de Lenguajes y Sistemas Informáticos,
Escuela Técnica Superior de Ingeniería Informática,
Avenida Reina Mercedes, s/n., 410121 Sevilla, Spain
`ajramirez@us.es`

1 Abstract

Automating business processes is one of the most recurrent topics in industries, independent of its digital orientation. Competitiveness pushes companies to deliver their products or services efficiently and effectively. Besides providing the appropriate value, they are required to do it faster and with higher quality. This agile context leads to automate *everything* that can be automated to keep the focus on the value while optimizing the processing times, errors, and process performance, in general [9].

Human beings have historically suffered various industrial revolutions that transformed the way of working, producing, and thinking. Although resistance to change has always appeared, they ended up being adopted by companies and people to avoid inevitable obsolescence [11]. The irruption of Robotic Process Automation (RPA) in the area of business process automation seems to have laid the seeds for a new revolution of administrative digital work [3].

RPA is a software paradigm that enables software machines (also referred as robots) to interact with information systems through their user interfaces (UIs) in a process-oriented way. Freeing humans from repetitive and mundane work is its main mantra. It started receiving increasing interest in the last decade and has become the fastest-growing enterprise software market in the last years [2]. After an initial hype of unfulfilled promises, RPA keeps a significant traction [12]. Nonetheless, some companies still fail when trying to incorporate RPA in their projects.

This paper serves as a discussion on, first, how to frame RPA in the existing Business Process Management (BPM) paradigm (cf. Sect. 1.1). And second, it deals with its natural evolution to a wider automation technology across the entire organization: *Hyperautomation* (cf. Sect. 1.2).

1.1 Framing RPA in BPM

Nowadays, a plethora of tools is available in the application landscape under the umbrella of RPA. However, their application scopes are wide, ranging from

This research has been supported by the Spanish Ministry of Science, Innovation and Universities under the NICO project (PID2019-105455GB-C31).

J. González Enríquez et al. (Eds.): BPM 2021, LNBIP 428, pp. 3–6, 2021.
https://doi.org/10.1007/978-3-030-85867-4_1

simple UI scripting tools (e.g., UI.Vision[1] or RobotFramework[2]) to comprehensive systems that enable the development, deployment, and control of farms of robots (e.g., UiPath[3] or Robocorp[4]).

This situation creates uncertainty in companies when deciding what to use for their use cases that typically leads to failed projects [4]. On the one hand, when neither scaling nor a central government of robots is required, UI scripting tools could do the job at a fraction of the cost when compared to mature RPA solutions, which use to be disproportionate in simple contexts. On the other hand, the hype created around RPA pushes some companies to use the technology to the detriment of other more suitable solutions which would deliver more outstanding performance. For example, *utilizing* RPA to automate UIs even though the API is exploitable leads to unnecessary inefficiencies and high resource consumption.

Even when the project (i.e., undesired contexts where no other automation alternative would work) fulfills the suitability criteria for RPA, companies may miss a threat analysis of the solution. In case that RPA is applied as a long-term solution (e.g., in legacy systems that cannot be changed), it becomes highly dependant on the UI of the base system. Therefore, monitoring or continuous testing is required to anticipate errors [6]. In turn, if RPA is applied as a short-term solution (e.g., rapid solution without investing in a deep integration), its end-of-life should be defined and control. Otherwise, it will become a technical debt in the team that has to do the maintenance [8].

The future shape of the RPA technology is uncertain since mature RPA vendors provide some features that overlap with those traditionally existing in the BPM tools, e.g., process modeling, orchestration, and monitoring. Nonetheless, while the *RPA-centric* solutions focus on fine-grained tasks, *BPM-centric* solutions support rather more complex and sophisticated integrations. What is more, this uncertainty is increased by the different market movements in both (1) delivering more BPM features by RPA vendors or (2) acquiring RPA solutions by BPM vendors[5].

What is clear is that both paradigms are part of a new *big thing* that enables the automation of a broader range of processes end-to-end. Independently on how they integrate, industry-grade solutions for RPA may support robot developers and robot operators/maintainers in a DevOps continuous cycle. In the development field, besides just creating and executing robots, additional features are necessary, like supporting identifying candidate processes to robotize, controlling the version of the robots, evolving them, or managing test suits in controlled environments. Regarding the operation field, besides the deployment

---

[1] https://ui.vision.

[2] https://robotframework.org.

[3] https://www.uipath.com.

[4] https://www.robocorp.com.

[5] As an example, in 2020 Appian acquired Jidoka RPA solution https://appian.com/resources/newsroom/press-releases/2020/appian-acquires-robotic-process-automation-rpa-company.html.

of the robots in the execution environments, this role must be supported with, for example, scaling and descaling mechanisms, balancing the workload of the robot queues, or alerting rules to control the correct performance. These requirements become even more challenging when we consider the participation of the human in the process. This is highly relevant in the automation with RPA since the automated processes here are typically those which were previously on the human side. As the automation does not happen like a big bang but through iterations [7], methods are required which consider the human in the process and that the work gradually shifts from the human side to the automatic/robotic side. Although robots have an initial relevant role in this shifting, the eventual automation solutions may use other more sound and resilient automation technology.

1.2   The Era of Hyperautomation

Hyperautomation is more than just a fancy word. It has been coined to combine BPM, AI, RPA, and any other technology that may help conduct human duties in an automatic way within organizations. Not surprisingly, Gartner identifies this technology as the number one trend in 2020[6].

While RPA scope still requires standardization, hyperautomation gives a name to this continuous effort to try to automate *everything* that can be automated. Similar to BPM and RPA paradigms, hyperautomation requires methods to ease its adoption. Here, the separation of duties and decoupling of each combined technology needs to be guarantee to allow their individual evolution. In the same way, streamlining the incorporation and coordination of different technologies within the available automation toolset is a must. Beyond processes and tasks, this technology aims at the organization's scope and, thus, new or adapted measures or KPIs are required to assess the automation level of the organization after each hyperautomation iteration. As already demonstrated in many similar contexts, process mining stands as a suitable technology to automated this assessment as well as to accelerate the discovery of potential automation alternatives, existing inefficiencies, etc. [5].

This shake to the whole organization will need to be addressed from different perspectives besides the DevOps one. From a strategic point of view, organizations need to reorder their priorities, rethink the management of their risks and resources, and, in summary, envision a future company that will require more technology, innovation, and smart minds with far less mundane and repetitive work. From a technological point of view, organizations need to agile the technology acquisition and mastering, enabling fast knowledge sharing and collaboration from different units or departments from both business and IT levels. In the center of this organization transformation is the human who, on the one hand, will suffer automation initiatives at higher rates than before, which may generate adverse reactions if they neglect to estimate the human impact of the

---

[6] https://www.gartner.com/smarterwithgartner/gartner-top-10-strategic-technology-trends-for-2020.

automation before its deployment [10]. On the other hand, human work habits will focus on more unique, cognitive, and valuable activities instead of batch-processing and simple ones.

One of the most determinant factors to successfully address all these dimensions of this journey to hyperautomation is to work on the skill developments at every tier of the company [1]. Continuous formations plans, knowledge transfer sessions, etc., are recommended in the area of automation. Current workers may benefit from existing literature and handbooks written for researchers and practitioners. In turn, a significant deficiency that needs to be faced is that the new generations—which typically came from universities and institutes—have access mainly to technical formation courses from vendors. However, both lectures and students lack comprehensive textbooks to get prepared for this new revolution called hyperautomation.

References

1. Now & Next: State of RPA (2021). https://www.automationanywhere.com/lp/now-and-next-rpa-report. Accessed 17 July 2021
2. Biscotti, F., Tornbohm, C., Bhullar, B., Miers, D.: Gartner Market Share Analysis: Robotic Process Automation, Worldwide, 2018, vol. G00385825. Gartner Research, Stamford (2019)
3. Fung, H.P.: Criteria, use cases and effects of information technology process automation (ITPA). Adv. Robot. Autom. **3**(3), 1–10 (2014)
4. Hindle, J., Lacity, M., Willcocks, L., Khan, S.: Robotic process automation: benchmarking the client experience. Tech. rep. (2018). https://www.knowledgecapitalpartners.com/research-and-publications/2018/2/5/rpa-benchmarking-the-client-experience-. Accessed 17 July 2021
5. Jimenez-Ramirez, A., Reijers, H.A., Barba, I., Del Valle, C.: A method to improve the early stages of the robotic process automation lifecycle. In: Giorgini, P., Weber, B. (eds.) CAiSE 2019. LNCS, vol. 11483, pp. 446–461. Springer, Cham (2019). https://doi.org/10.1007/978-3-030-21290-2_28
6. Jiménez-Ramérez, A., Chacón-Montero, J., Wojdynsky, T., González Enríquez, J.: Automated testing in robotic process automation projects. J. Softw. Evol. Process n/a(n/a), e2259 (2020). https://doi.org/10.1002/smr.2259. https://onlinelibrary.wiley.com/doi/abs/10.1002/smr.2259, e2259 smr.2259
7. Jiménez-Ramírez, A., Reijers, H.A., González Enríquez, J.: Human-computer interaction analysis for RPA support, pp. 169–186. De Gruyter Oldenbourg (2021)., https://doi.org/10.1515/9783110676693-009
8. Kampik, T., Hilton, P.: Towards social robotic process automation. In: SIAS Conference (2019)
9. Kirchmer, M., Franz, P.: Value-driven robotic process automation (RPA). In: Shishkov, B. (ed.) BMSD 2019. LNBIP, vol. 356, pp. 31–46. Springer, Cham (2019). https://doi.org/10.1007/978-3-030-24854-3_3
10. Parasuraman, R., Sheridan, T., Wickens, C.: A model for types and levels of human interaction with automation. IEEE Trans. Sys. Man. Cybernet. Part A Syst. Hum. **30**(3), 286–297 (2000). https://doi.org/10.1109/3468.844354
11. Émile Pouget: Sabotage. Charles H. Kerr & Company, Chicago (1913)
12. Taulli, T.: Future of RPA. In: The Robotic Process Automation Handbook, pp. 293–316. Apress, Berkeley (2020). https://doi.org/10.1007/978-1-4842-5729-6_13

A Framework of Cost Drivers for Robotic Process Automation Projects

Bernhard Axmann[1] , Harmoko Harmoko[1] ,
Lukas-Valentin Herm[2(✉)] , and Christian Janiesch[2,3]

[1] Technische Hochschule Ingolstadt, Ingolstadt, Germany
{bernhard.axmann,harmoko.harmoko}@thi.de
[2] Julius-Maximilians-Universität, Würzburg, Germany
{lukas-valentin.herm,
christian.janiesch}@uni-wuerzburg.de
[3] HAW Landshut, Landshut, Germany

Abstract. Robotic process automation is a technology to imitate human behavior when interacting with computers to perform digitized tasks manually, such as opening and closing applications, reading documents, entering data, and sending e-mails. As with any new technology, estimating the costs and break-even of robotic process automation projects is challenging. Currently, in practice, there are no dedicated guidelines for defining cost components in those projects that go beyond simple comparison with person-hours and salary cost. To address this gap, we review literature on the cost of robotic process automation projects to collect and structure those cost drivers that can be generalized. We categorize and prioritize them and derive a novel cost framework specifically for the cost estimation of robotic process automation projects. The framework comprises three cost calculation perspectives for three distinct project scopes hosting eleven cost drivers in the three categories development, investment, and operation. We illustrate the framework in a robotic process automation use case to demonstrate its usefulness.

Keywords: Robotic process automation · Literature review · Cost drivers · Framework

1 Introduction

Robotic process automation (RPA) has recently moved from the peak of exaggerated expectations into the trough of disillusionment in Gartner's hype cycle for legal and compliance technologies [1]. It indicates that business is better understanding "what RPA can do" and "what RPA cannot do" in its current form nearing the plateau of productivity.

Nevertheless, according to Syed et al. [2] measuring RPA cost and benefit realization still poses a challenge as there is no readily available metric system or even a common understanding of the benefits and cost of RPA. This is chiefly because RPA is a novel approach to IT automation and not directly comparable to traditional IT-related projects, developing a novel IT system, or implementing workflows using standalone

J. González Enríquez et al. (Eds.): BPM 2021, LNBIP 428, pp. 7–22, 2021.
https://doi.org/10.1007/978-3-030-85867-4_2

business process management (BPM) software to automate a task rather than automate the human worker [3].

In contrast, due to their simple structure, it is tempting to break down RPA projects into "wins" or "fails" at first glance. However, in reality, it is not nearly as trivial to understand the cost-benefit relation at work as not all cost drivers are apparent at all times. Hence, a reasonable first step to approach RPA project measurement is to focus on the cost incurred by the development, investment, and operation of RPA as – in contrast to benefits – costs can be better quantified and assessed early in projects [4].

While it may sound promising now to compile a comprehensive framework of RPA cost drivers, any approach for RPA has to align with the lightweight character of RPA projects, which does not always allow for complex organizational and budgetary considerations. Otherwise, hyperautomation, the rapid, business-driven realization of RPA software robots [5], would not scale.

In response, we propose an investigation into the cost drivers that may affect RPA while maintaining the stance that cost estimation for RPA projects needs to be reasonably straightforward to be effective given the circumstances of application. We formulate our research question accordingly:

RQ: *Which cost drivers are relevant for the cost estimation of RPA projects and how can we incorporate them in a framework applicable for different project scopes?*

Our contribution is threefold. First, we contribute a comprehensive review of the state-of-the-art of RPA project costing that reveals cost drivers that have been applied to the measurement RPA cases. Second, we offer a conceptualization of RPA projects in size and complexity as well as a differentiation of distinct ways of measuring cost, which together form a 3 × 3 costing matrix of RPA projects. Third, we suggest a comprehensive collection of cost drivers for RPA project cost estimation that can be situated in the costing matrix to provide guidance for RPA project of different scope, sizes, and extent.

The paper is structured as follows: In Sect. 2, we present the theoretical foundation for cost estimation metrics. Section 3 comprises the research design, including details on the conducted literature review. Section 4 introduces our novel framework for cost estimation, which we illustrate in Sect. 5 and discuss in Sect. 6. Lastly, in Sect. 7 we conclude with a summary and discussion.

2 Foundations for Cost Estimation in IT Automation Projects

2.1 Comparison of Cost in IT Automation Projects and RPA Projects

Traditional IT automation projects implementing enterprise software or BPM software are considered to be heavyweight implementation projects, as they require long-term and intensive design and testing within the already existing IT infrastructure [2, 6]. Similarly, these automation projects require application programming interfaces (API) to existing software to automate tasks across different applications. These APIs are not always available or sufficient and sometimes must be implemented first, before any automation can be approached [7]. In contrast, RPA represents a lightweight automation approach as it only focuses on the pre-existing presentation layers of (enterprise) software enabling rapid development and rollout [3]. Compared to traditional automation projects, RPA aims to imitate human behavior rather than automate tasks within the IT backend. This anthropomorphic characteristic allows business users, without extensive programming knowledge, to develop RPA robots on their own that mimic themselves [8].

Hence, it is evident that these characteristics also impact the resulting costs of implementations [2]. While automation projects for enterprise software mainly produce high costs due to their lengthy and labor-intensive customization [9], RPA is often considered as a bridging technology that enables rapid automation until backend integration is financially and organizationally feasible [3]. Thus, in traditional automation projects, costs usually arise before and during project implementation [2, 6]. In contrast, in RPA software robots that only access the presentation layer of legacy software are prone to errors as adjustments to the user interface (UI) can cause the software robot to become inoperable [7]. This results in long-term costs for operating and maintaining RPA robots [10]. Consequently, the nature and timing of how cost drivers occur may vary compared to traditional automation projects.

However, while various cost drivers that occur during and after a traditional automation project are well researched [6], the lack of holistic and scientific studies focusing on cost drivers within RPA projects becomes apparent [2]. To measure RPA costs more accurately and appropriately, they need to be systematized in a structured fashion.

2.2 Dimensions of Measurement Metrics

To measure the success of an IT automation project, companies must compare benefits and costs of such implementations in the long and short term. While calculating benefits can be challenging as they can arise as both quantitative and qualitative factors, calculating costs usually relies on quantitative drivers [11]. In practice, the success evaluation is done by applying several key performance indicators focusing on different aspects and time periods [11–13].

Theory and practice distinguish these key performance indicators in *absolute* and *relative metrics* to differentiate between the overall success and the effectiveness of a project [11, 14–19]. The effectiveness is measured with the return on investment (ROI) [16–18, 20, 21]. These two metrics are supplemented with the metric *time to ROI* to enable alignment with a company's strategic plans. Time to ROI is the time required for benefits to equal costs [18, 21]. Consequently, measuring success in various RPA implementation projects requires taking these key performance indicators into account.

3   Research Methodology

Our research is based on a structured literature review, which we used to synthesize existing scientific considerations on the costing of RPA projects. Further, we extended our structured literature review through the integration of practical contributions to provide a comprehensive overview of research and practice. In the following we describe a comprehensive overview of our procedure and a meta-synthesis of the findings.

Procedure of Structured Literature Review. To ensure scientific rigor, we conducted a structured literature review according to vom Brocke et al. [22]. Since RPA is intersecting many research fields, we queried multiple databases. We focused on the computer science related databases ACM Digital Library and IEEE Xplore. Further, we queried the information systems related databases Science Direct and AIS eLibrary. Lastly, we included the database SpringerLink, for contributions from multidisciplinary research fields. Thereby, we use following search query: *"RPA OR 'robotic process automation'"*. In doing so, we intentionally kept the search query generic to avoid excluding articles that indirectly discuss various cost drivers of RPA. Further, due to the novelty of the subject, we did not restrict our research results to any form of outlet rankings. Following this strategy, we found 1,522 academic contributions dealing with the topic of RPA. Through abstract, title, and keyword analysis, followed by full-text analysis, as well as forward and backward search of the remaining contributions, we found only $n = 8$ academic contributions that were relevant to our cause. We have classified contributions as relevant that describe and discuss the cost drivers in detail, rather than just naming them. Figure 1 provides an overview of the process.

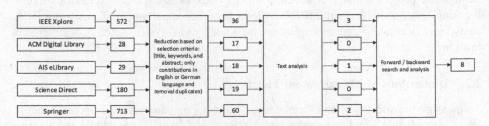

Fig. 1. Results of the literature review according to vom Brocke et al. [22]

Extension of Structured Literature Review. Since we found only $n = 8$ contributions dealing with cost drivers, we extended our results by integrating practice-oriented and, thus, non-peer-reviewed contributions. To do so, we used the search query from our structured literature review and applied it to Google search. Here we have focused on reports and white papers from consulting firms and software development companies, describing real-world RPA implementations. The prioritization of the result presentation was done by the Google search based on the search query. We followed Aldiabat et al. [23] and terminated our search when we noticed content saturation. To ensure an adequate relation with academic papers, settled for a total of $n = 8$ effectually distinct practice contributions dealing with cost drivers. As a result, we used $n = 16$ contributions from academia and practice to derive our framework.

Analysis Procedure. We followed vom Brocke et al. [22]'s recommendation and subdivided the findings into units of similar content to synthesize previous research. Thus, we investigated all contributions to derive different costs drivers. Then, we compared these cost drivers, grouped similar drivers, and revisited all contributions based on these findings. As a result, we derived $n = 11$ different cost drivers from both types of literature. Since, the classification was performed by a single coder, we followed the recommendations of Fleiss' Kappa statistics [24] and performed a blindfolded classification with a second coder, to ensure the quality of our results. The comparison of the classifications yielded a so-called "Excellent" result *(k: 0.87)*.

Meta-Analysis. In Fig. 2, we present a meta-analysis of the derived cost drivers. On the left side, we show the distribution of cost drivers for each year, as well as the overall proportional distribution on the right side.

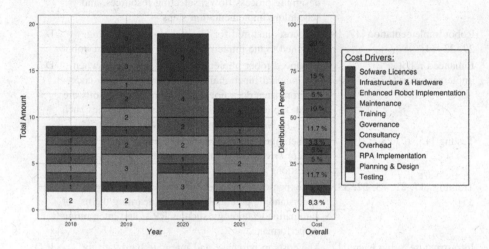

Fig. 2. Meta-synthesis of literature review

Looking at the results, it is noticeable that in the early stages of RPA research (2018), the number of reported cost drivers ($n = 8$) is relatively low. In contrast, since 2019, there has been an increase in contributions presenting different cost drivers. While cost drivers such as software licenses ($n = 12$, $\approx 20\%$), infrastructure and hardware ($n = 9$, $\approx 15\%$), training ($n = 7$, $\approx 12\%$), and RPA implementation ($n = 7$, $\approx 12\%$) were mentioned most frequently, cost drivers such as overhead ($n = 3$, $\approx 5\%$) or governance ($n = 2$, $\approx 3\%$) were mentioned only a few times.

4   Cost Estimation in Robotic Process Automation Projects

4.1   Cost Drivers in RPA Projects

RPA projects consist of three general phases: initialization, implementation, and scaling [3, 25]. Each stage has consequences to the costs of a project. The benefits of RPA, such as the efficiency and accuracy of the process [26], are offset by the costs, making RPA investments profitable [27, 28]. The cost components of RPA depend on the number and types of software robots as well as the scale and complexity of the process [29, 30].

Based on our literature review, we identified eleven cost drivers, which we grouped into: development (D), investment (I), and operation (O), as shown in Table 1.

Table 1. Cost drivers in RPA projects

Costs	Description	Group
Consultancy [17, 25, 31, 32]	External experts supporting the limited resources and skills of a company implementing RPA	D
Planning & design [15, 17, 25, 33]	Initial project-management-related activities including assessing potential processes, (re-) designing process flows, selecting resources, and scheduling implementation steps	D
Robot implementation [17, 18, 27, 32–34]	The costs incurred for internal employees being assigned to the implementation of the software robot	D
Enhanced robot implementation [17, 28, 32]	Enhanced robot implementation is necessary when the standard functionality of typical RPA low-code environments does not suffice to realize the software robot, typically to realize intelligent capabilities such as OCR, NLP, or image recognition	D
Testing [15, 17, 28, 32]	All activities to observe, record, and evaluate the system or component under specific operating conditions	D
Training [17, 21, 25, 32, 33, 35]	All expenses to train technical and business staff in understanding potential processes, learning the use of automation technology such as RPA, and interpreting its performance	D
Infrastructure & hardware [17, 18, 27, 28, 31–33, 35]	The costs to purchase and integrate hardware for hosting, implementing, and running software robots. That is primarily (virtual) servers and desktops. These can be bought or rented	I, O
Software licenses [17, 18, 21, 27, 28, 32–39]	Software is necessary to run servers, desktops, develop and software robots. Additional licenses may also be necessary to access legacy software. Licenses can be bought or subscribed	I, O
Governance [35]	Costs associated with managing the new organizational structure after introduction software robots. For example, data governance, infrastructure or governance & IT governance	O
Maintenance [15, 21, 25, 28, 32, 33, 35]	All expenses related to continuous maintenance of software robots. E.g., software, hardware, updates, and human resources for exception handling, maintenance and improvement	O
Overhead [27, 32, 39]	Costs incurred to support automation process which do not directly involve RPA investment. For example, HR, Finance, IT administration, rent, utilities, insurance, office supplies, as well as accounting and legal expense	O

Legend of Cost Group: D: Development, I: Investment, O: Operation.

Development Cost. Development cost are one-time costs, which summarize internal and external personnel costs. In IT projects, these costs can exceed other costs groups many times over and have to be considered over multiple periods. Implementation costs involve expenditures related to preparing, installing, configuring, and deploying RPA solutions in an organization [35]. It is generally agreed that the cost of RPA implementations is significantly lower than traditional IT development projects ranging from small-scale custom development to enterprise software introduction although all require special knowledge for their development [38]. Typical development costs are planning & design, consultancy, training, RPA development, testing, and enhanced development cost.

Investment Cost. Investment cost are one-time activities as well that typically generate costs early in a project. However, investments are not limited to the initialization phase of the project, but they can occur in any phase. When a project scales, such as adding unattended software robots or adding cognitive capabilities (e.g., object character recognition (OCR), natural language processing (NLP), or image recognition), the company must invest in new software or hardware. Typical investment costs are software licenses and infrastructure & hardware.

Operation Cost. Operation cost are all the expenses to run the respective software within an organization. This entails that the costs will continue to exist as long as the software robots operate [27]. Synonyms are running costs or ongoing costs. The operating costs consist of rented software licenses and infrastructure, maintenance, overhead, and governance costs.

4.2 Systematization of Cost Estimation in RPA Projects

When creating a measurement framework, there are two inherent conflicts. One conflict is inherent to the information content: Should the information be presented and considered as accurate as possible or as easily accessible as possible? The second conflict is about choosing the right RPA project and thus the proper prioritization: Should it be implemented as quickly as possible with high efficiency, or should it have a significant impact?

The solution to these conflicts is to offer not one metric but a set of metrics. The 3×3 matrix shown in Fig. 3 solves these two conflicts with a systematic effort- and situation-oriented approach suitable for RPA.

The metrics for information content are divided into the *project scopes* of a *single RPA robot*, *multiple RPA robots*, and *institutionalized use of RPA*. When using RPA for a single robot, there is no need for a detailed cost-benefit analysis. There is only the need to provide a quick assessment whether the automation project can make a difference. However, the effort and the dedication to implement the very first robot must be adequate to set a good example. However, when setting up multiple software robots further factors (maintenance, enhanced robot implementation, testing, training, infrastructure & hardware) have to be considered. Moreover, institutionalizing RPA in an organization as an automation paradigm requires further oversight that must be incorporated in cost measurements (overhead, governance, planning & design). A more detailed description follows and is also made transparent in Fig. 4.

According to Herm et al. [3], generally recommended steps for implementing RPA projects are identification, alignment, screening, evaluation of business case, process selection, RPA software selection, proof of concept, and RPA rollout. The coordination effort and interdependencies increase naturally when several RPA projects are conducted at the same time. Accordingly, the need to present the costs in more detail increases. Nevertheless, the effort for evaluation should still be reasonable and the framework not overly complex for implementation projects of single software robots resulting in needs for the scope of a *single software robot*.

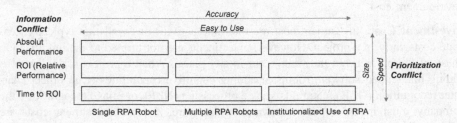

Fig. 3. 3 × 3 matrix for cost estimation in RPA projects

The steps involved in implementing *multiple software robots* are basically the same as for one RPA project. Only the selection of the RPA software can be omitted, since it is recommended to work only with one RPA software. However, the complexity to manage multiple projects at the time rises.

When RPA is introduced for *institutionalized use* into a company, the complexity and effort increases even further with the additional benefits of increased efficiency and oversight. In addition to the project-specific efforts already described, efforts sustainable integration into the organization and scaling must be considered [3]. Respective RPA support processes are management support, change management, IT integration, and governance to integrate RPA sustainably. Scaling activities include the efforts to run, grow, or eventually retire RPA projects [3]. These activities require a detailed analysis of the costs as they have to be contrasted to heavyweight integration alternatives. High accuracy of the analysis is more important than a fast analysis with low effort.

Regardless of the *project scope*, the *measurement dimension* is essential when measuring RPA projects. That is, it is essential to decide whether the goal is to achieve success quickly or to achieve great success. Statements about the magnitude and size of the success are given by *absolute metrics*. Statements about the speed and effectiveness of success are given by *relative metrics*. Both are important for measurement and must be weighed against each other when managing projects. For first RPA projects and the pilots, it makes more sense to select projects with faster success to prove the technology's benefits and identify strengths and weaknesses of the technology early on [3]. It is to be expected that these early projects tend to be significantly smaller in the absolute dimension. These measurements can be supplemented by the *time to ROI*, as this metric illustrates the effectiveness over time and indicates when the invested costs break-even. To differentiate the three dimensions, the following description provides more details:

- **Absolute metrics** quantify the significance and size of the success for the company.
- **Relative metrics** quantify the project success in comparison to existing projects in the company in terms of speed and effectiveness.
- **Time to ROI** helps to synchronize the projects with the strategic plan of the company and is also a relative metric.

To evaluate the success once benefits are known, a classic cost-benefit analysis can be used.

4.3   A Framework of Cost Estimation with Varying Project Scopes

In the following, we illustrate a framework based on the extended literature review and the theoretical considerations above. Additionally, we differentiate all cost drivers according to the project scope of occurrence (single, multiple, institutionalization), cost group (D, I, O), and the type of occurrence (optional, mandatory). See Fig. 4 for an overview.

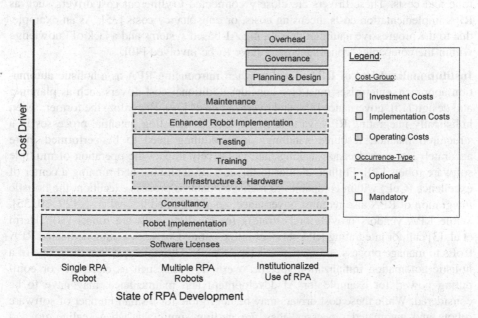

Fig. 4. A framework for cost estimation of RPA projects

Single RPA Robot. Within a single software robot implementation, cost estimation should be quickly calculable. Hence, only the direct cost drivers related to software licensing costs [21], robot implementation [17], and consultancy costs [25] shall be considered, whereby consulting and licensing costs may even be optional during this stage of development [30]. Looking at software licensing costs, RPA vendors often provide trial licenses or community editions with a limited feature set [30]. Similarly, due to the low-code nature of RPA, there may not be a need for dedicated programmers or external consultants. As a result, business users can develop initial RPA robots themselves and thus only generate implementation costs without the need for external cost or additional RPA developers [29, 30]. Finally, infrastructure costs are negligible at this stage, as a software robot can run attend on any user's desktop [30, 38].

Multiple RPA Robots. In contrast, conducting projects to realize multiple software robots at the same time also entails further costs. Costs such as training [25], testing [15], and maintenance [35] are mandatory to consider, while infrastructure costs [21] and additional implementation costs [35] may incur. Based on our research, we noticed that the need for training and teaching additional employees is necessary to ensure a general acceptance regarding the integration of a software robot due to its anthropomorphic nature [29, 30]. Similarly, when applying multiple software robots for a business process, testing and maintaining these robots have to receive more awareness and diligence to avoid side effects [15, 30]. In contrast, the integration of additional software robots for already automated process, for example, when scaling up execution volume [21, 29] or integrating intelligent features such as NLP for the automation of further process variations [35], will – when applied – result in additional development time and costs. These drivers are closely connected to different cost drivers such as RPA implementation costs, licensing costs, or consultancy costs [35]. As an example, due to the progressive nature of developing AI-based systems and a lack of knowledge within the companies, consultants may have to be involved [40].

Institutionalized Use of RPA. Finally, when introducing RPA as a holistic automation approach, enterprises must also consider additional cost drivers such as planning and design [15], governance [35], and overhead costs [39]. Regarding the former, when holistically integrating RPA, various steps, such as selecting potential processes in a structured manner, resource handling, or scheduling need to be performed more accurately, as poor decision-making may negatively impact the operation of multiple software robots [30]. Further, similar to the cost of deploying and running a center of excellence (CoE) within companies to manage business process execution, the holistic integration of RPA also requires governance, resulting in additional costs [29, 30, 35]. While RPA vendors provide orchestrators for handling software robots [30], Herm et al. [3] call for integrating RPA services into existing CoEs or creating dedicated RPA CoEs to manage process automation at a larger scale. Ultimately, within the scope of a holistic automation initialization specific overhead costs such as electricity or computing power, for example for AI development and maintenance, may have to be considered. While these cost drivers may be negligible for a small number of software robots and automated processes, they are getting significant when scaling up and institutionalizing IT automation with RPA [29, 30].

5 Illustrative Use Case[1]

Imagine a large multinational conglomerate company or a medium-sized or large enterprise that requires data aggregation across multiple sources into a management dashboard. This data is often not convenient or accessible via a structured extract, transform and load process. The case aggravates for data from cloud applications that only provide front-end access.

Now to populate said management dashboard with the required information, an employee, a temporary worker, or an intern collects this data via copy and paste from the various application screens and formats, aggregates, and integrates the data into the target application. Each week or month this tedious task takes a lot of time and is prone to human errors of pasting data in the wrong field or mistyping a value.

Trialing a *single RPA robot* in any company size and setup should be a straightforward exercise with minimal costing requirements. Here, the company uses its own resources to implement the software robot for the most tedious data aggregation activities using a free trial license running on the employee's desktop overnight. Hence, cost estimation should solely be based on the cost of implementing the software robot and held against the benefits the company experiences in relative metrics.

Once the software robot is in operation, more data aggregation tasks are sought to be automated and data entry tasks could also be included. This requires further software robots that will work in parallel. Hence, cost estimation needs to consider costs for *RPA software robots*. That is, testing, training, maintenance becomes more of a structured activity and cannot be subsumed under implementation cost. Further, appropriate license agreements need to be made in a structured fashion as they may lead to a vendor lock-in. Moreover, the software robots require more resources to run 24/7 and therefore infrastructure must be (internally) rented, which incurs cost. Lastly, there is data that standard robots cannot copy as they originate from scanned letters. So far, a human worker had to perform these tasks. Now, advanced OCR functionality shall be implemented by external consultants working with internal developers. Altogether, this requires a more detailed calculation of the cost due to the scope not only in relative but also in absolute terms.

After several months of operation, the company is very happy with the automation approach of RPA and intends to *institutionalize the use of RPA*. That is, they establish RPA as a bridging technology before heavyweight projects can create value and a cost-effective alternative for data aggregation and migration, especially from cloud services. This is more than the sum of its parts (i.e., its robots). Hence, to establish a structured prioritization pipeline of multiple projects, provide costing templates for new projects, and manage multiple projects at the same time, an RPA CoE must be established and staffed for planning and design activities as well as to monitor and govern RPA proliferation in the company. As soon as this army of software robots requires a noticeable amount of utilities as well as occupies full-time personnel, overheads for

---

[1] The use case is a synthesis of multiple workshops, we conducted with companies actively using RPA and combines their actual processes, project goals, and considerations for future improvements in one hypothetical use case.

electricity, compute power, etc. as well as office space may need to be included in the estimation of RPA cost. This cost will be measured in relative, absolute, and especially for bridging use cases in terms of time to ROI.

6 Discussion

Our cost estimation framework in RPA projects can be applied based on three project scopes with up to eleven distinct mandatory or optional cost drivers using a 3×3 costing matrix. It is a holistic approach that is suitable for single RPA robots to test out the technology initially, but it is also applicable for the institutionalized use of RPA as an automation paradigm in any organization. Our results come with several theoretical and practical implications that we discuss in the following:

Theoretical Implications. Following Kohli and Grover [41] our research contributes towards manifesting, when an RPA automation project can add value to companies. While IT creates value through many different aspects such as competitive advantage or increasing operational effectiveness [41] our research focuses on the investigation of monetary aspects namely cost drivers. Even though the results of the academic literature review were limited, our survey of cost drivers is the first comprehensive analysis of this emerging topic. Our consideration of practice reports does not only enrich and justify our findings, but it extends it with current observations from the field that we scrutinized theoretically. While our dimensions to structure cost drivers for RPA projects are not unique to RPA, and none of the concepts is necessarily unheard of, their combination provides a novel systematization of costs in RPA projects that can be readily applied in multiple projects scopes with multiple perspectives on cost estimation. Therefore, it provides a novel lens through which one can consider cost estimation in RPA projects. It enables straightforward measurements with minimal overhead that are of immediate practical value as well as comprehensive considerations of the extended costs of widespread and institutionalized automation with RPA. Our framework is extensible. That is, the emergence of novel cost drivers due to the evolution from symbolic RPA to intelligent RPA [40] can be incorporated naturally. It is grounded in theory but designed for practical application.

Practical Implications. First and foremost, the framework offers a practical guideline that aims to be useful for practice and extends beyond the simplistic calculation of *"salary times working hours"* prevalent in contemporary RPA projects. With the 3×3 costing matrix, cost estimation can be contextualized to fit a range of project scopes. It is simple enough to be of immediate pragmatic use to structure costing in early RPA projects, but it is also a comprehensive guideline to consult when embarking or continuing one's automation journey with RPA. That is, a company's cost estimation using the framework can grow as the projects grow from the first software robot to the institutionalized use of this technology. However, these cost drivers will change as companies begin to incorporate other automation technologies such as self-learning robots [40] that can maintain themselves.

Peculiarities of RPA Implementation Costs. While the cost drivers presented in Table 1 are derived from RPA-based literature, we found similarities to cost drivers from traditional automation projects. In this context, cost drivers such as consulting, planning & design or testing are also relevant in these types of projects [2, 6]. However, the occurrence time differs, since many stages in traditional projects are performed before the rollout, compared to the rapid development and continuous maintaining behavior of RPA projects [3]. Also, when comparing RPA implementations with automation techniques that use artificial intelligence (AI), e.g., cognitive automation, many similarities and differences become apparent. For example, while AI-based systems must deal with drifts in the data, resulting in continuous adjustments [40], RPA robots have to be maintained when the presentation layer of software is changed. Further, cost driver such as consultancy, planning & design, testing, training, or hardware & infrastructure, also apply to these automation projects [42] However similar to traditional implementation projects, most of these cost drivers are primarily incurred before the actual rollout, as AI-based systems require a lot of data collection and heavyweight training, especially when it comes to neural networks [43].

Limitations. Our work is not without limitations as the focus of our work was solely on the cost perspective. We did not yet systematize benefits of RPA, which would be necessary for a comprehensive cost-benefit calculation. Further, we did not survey costing for just any type of automation project and, thus, we may have missed a suitable metric. Thus, we were not able to quantify the change in costs for the various drivers when scaling up the project scopes. In addition, we stayed on the level of cost drivers and did not analyze individual cost factors and aggregated them to concrete formulae. Lastly, we used the scenario technique and theoretical arguments to assess our cost framework and did not perform naturalistic evaluation workshops or interviews with practice. These are subject to future research.

7 Conclusion and Outlook

To the best of our knowledge, a comprehensive collection and structuring of cost drivers for RPA has not been undertaken and constitutes a research gap and practical problem for many companies. In response, we categorized and prioritized cost drivers for RPA projects and derived a novel cost framework specifically for the cost estimation in such projects. We illustrated the framework's usefulness and discussed the framework's implication for theory and practice. Our implication entails that the framework provides a novel lens to analyze costs in RPA projects and assists practice in objectively budgeting and reviewing costs in RPA projects. To counter our limitations, future research will need to review and adapt cost formulae to suit RPA, apply the framework to real-life cases, and revisit costing in other automation domains to assess the transferability of results.

References

1. van der Meulen, R.: 4 Key Trends in the Gartner Hype Cycle for Legal and Compliance Technologies. Legal and Compliance, 20 September 2020. https://www.gartner.com/smarterwithgartner/4-key-trends-in-the-gartner-hype-cycle-for-legal-and-compliance-technologies-2020/. Accessed 14 May 2021
2. Syed, R., et al.: Robotic process automation: contemporary themes and challenges. Comput. Ind. **115**, 103162 (2020)
3. Herm, L.-V., et al.: A consolidated framework for implementing robotic process automation projects. In: Fahland, D., Ghidini, C., Becker, J., Dumas, M. (eds.) BPM 2020. LNCS, vol. 12168, pp. 471–488. Springer, Cham (2020). https://doi.org/10.1007/978-3-030-58666-9_27
4. All Answers Ltd.: Factors Affecting Cost Estimation on Project Cost Management in Construction Firms, 1 November 2018. https://ukdiss.com/examples/factors-affecting-cost-estimation.php#citethis. Accessed 20 May 2021
5. Ray, S., Tornbohm, C., Kerremans, M., Miers, D.: Move beyond RPA to deliver hyperautomation. EA and technology innovation leaders are often challenged to create a strategy that can capitalize on DigitalOps competencies and tools (2019). https://www.gartner.com/en/doc/433853-move-beyond-rpa-to-deliver-hyperautomation. Accessed 17 May 2021
6. Dumas, M., La Rosa, M., Mendling, J., Reijers, H.A.: Fundamentals of Business Process Management. Springer, Heidelberg (2018). https://doi.org/10.1007/978-3-662-56509-4
7. Elragal, A., Haddara, M.: The use of experts panels in ERP cost estimation research. In: Quintela Varajão, J.E., Cruz-Cunha, M.M., Putnik, G.D., Trigo, A. (eds.) CENTERIS 2010. CCIS, vol. 110, pp. 97–108. Springer, Heidelberg (2010). https://doi.org/10.1007/978-3-642-16419-4_10
8. van der Aalst, W.M.P., Bichler, M., Heinzl, A.: Robotic process automation. Bus. Inf. Syst. Eng. **60**(4), 269–272 (2018). https://doi.org/10.1007/s12599-018-0542-4
9. Daneva, M., Wieringa, R.: Cost estimation for cross-organizational ERP projects: research perspectives. Softw. Qual. J. **16**(3), 459–481 (2008). https://doi.org/10.1007/s11219-008-9045-8
10. Noppen, P., Beerepoot, I., van de Weerd, I., Jonker, M., Reijers, H.A.: How to keep RPA maintainable? In: Fahland, D., Ghidini, C., Becker, J., Dumas, M. (eds.) BPM 2020. LNCS, vol. 12168, pp. 453–470. Springer, Cham (2020). https://doi.org/10.1007/978-3-030-58666-9_26
11. Blaha, C., Graf, A., Heimel, J., Meier, T., Niedermayr, R., Schumacher, W., Hubert, T.: Controlling Process KPIs. A Guideline for Measuring Performance in Controlling Processes, p. 16 (2012). https://www.igc-controlling.org/fileadmin/downloads/Standards/ControllingProcessKPIs.pdf. Accessed 21 May 2021
12. Cruz Villazón, C., Sastoque Pinilla, L., Otegi Olaso, J.R., Toledo Gandarias, N., López de Lacalle, N.: Identification of key performance indicators in project-based organisations through the lean approach. Sustainability **12**, 5977 (2020)
13. McGlynn, L.: Quantitative Versus Qualitative KPIs, 25 June 2015. https://www.laverymcglynn.co.uk/blog/news/quantitative-versus-qualitative-kpis. Accessed 20 May 2021
14. Horváth, P., Gleich, R., Seiter, M.: Controlling, p. 307. Verlag Franz Vahlen/ProQuest, München/Ann Arbor, Michigan (2020)
15. Ma, Y., Lin, D., Chen, S., Chu, H., Chen, J.: System design and development for robotic process automation. In: IEEE International Conference on Smart Cloud (SmartCloud), Tokyo, pp. 187–189 (2019)

16. White, D.C.: Calculating ROI for Automation Projects (2007). https://www.emerson.com/documents/automation/white-paper-calculating-roi-for-automation-projects-deltav-en-40896.pdf. Accessed 10 May 2021
17. UIPath: Performing the Cost Benefit Analysis, 26 April 2021. https://docs.uipath.com/automation-hub/docs/performing-the-idea-cost-benefit-analysis#cost-estimates. Accessed 16 May 2021
18. van den Oever, B.: Method for estimating the impact of Robotic Process Automation implementations on business processes, 1 June 2020. http://dspace.library.uu.nl/handle/1874/397880. Accessed 15 May 2021
19. Butler, K.M.: Estimating the economic benefits of DFT. IEEE Des. Test Comput. **16**, 71–79 (1999)
20. Fernández, J.F.G., Márquez, A.C.: Control and knowledge management system. In: Gómez Fernández, J.F., Márquez, A.C. (eds.) Maintenance Management in Network Utilities, pp. 299–329. Springer, London (2012). https://doi.org/10.1007/978-1-4471-2757-4_12
21. Automation Anywhere: Business Analyst Enterprise (v11) (2021). https://university.automationanywhere.com/training/rpa-learning-trails/business-analyst/. Accessed 15 May 2021
22. Vom Brocke, J., Simons, A., Riemer, K., Niehaves, B., Plattfaut, R., Cleven, A.: Standing on the shoulders of giants: challenges and recommendations of literature search in information systems research. Commun. Assoc. Inf. Syst. **37**(1), 9, 205–224 (2015)
23. Aldiabat, K.M., Le Navenec, C.L.: Data saturation: the mysterious step in grounded theory methodology. Qual. Rep. **23**(1), 245–261 (2018)
24. Fleiss, J.L.: Measuring nominal scale agreement among many raters. Psychol. Bull. **76**, 378–382 (1971)
25. Hallikainen, P., Bekkhus, R., Pan, S.L.: How OpusCapita used internal RPA capabilities to offer services to clients. MIS Q. Exec. **17**(1), 41–52 (2018)
26. Vitharanage, I.M.D., Bandara, W., Syed, R., Toman, D.: An empirically supported conceptualisation of robotic process automation (RPA) benefits. In: European Conference on Information System (2020)
27. Institute for Robotic Process Automation and Artificial Intelligence (IRPA AI): Understanding RPA ROI (2019). https://irpaai.com/understanding-rpa-roi-sponsored-measure-important-2/. Accessed 15 May 2021
28. Genpact, Inc.: From robotic process automation to intelligent automation. Six best practices to delivering value throughout the automation journey (2018). https://www.genpact.com/downloadable-content/insight/the-evolution-from-robotic-process-automation-to-intelligent-automation.pdf. Accessed 12 May 2021
29. Taulli, T.: The Robotic Process Automation Handbook. A Guide to Implementing RPA Systems. Apress, Berkeley (2020)
30. Tripathi, A.M.: Learning Robotic Process Automation. Create Software Robots and Automate Business Processes with the Leading RPA tool (UiPath). Packt Publishing, Birmingham (2018)
31. Costin, B.V., Anca, T., Dorian, C.: Enterprise resource planning for robotic process automation in big companies. A case study. In: 24th International Conference on System Theory, Control and Computing (ICSTCC), pp. 106–111. IEEE (2020)
32. Willcocks, L., Hindle, J., Lacity, M.: Keys to RPA Success (2019). https://www.blueprism.com/uploads/resources/whitepapers/KCP_Report_Change_Management_Final.pdf. Accessed 15 May 2021
33. Almog, D., Bezobrazova, Y., Zlotova, V., Kadosh, G.: Robotic Process Automation. Total Cost Automation (2020). https://go.kryonsystems.com/kryon-and-ey-rpa-whitepaper. Accessed 15 May 2021

34. Enriquez, J.G., Jimenez-Ramirez, A., Dominguez-Mayo, F.J., Garcia-Garcia, J.A.: Robotic process automation: a scientific and industrial systematic mapping study. IEEE Access **8**, 39113–39129 (2020)
35. Deloitte Development LLC & Blue Prism: Calculating real Calculating real ROI on intelligent automation (IA) (2020). https://www2.deloitte.com/content/dam/Deloitte/us/Documents/technology-media-telecommunications/blue-prism-white-paper-final.pdf. Accessed 15 May 2021
36. Benkalai, I., Seguin, S., Tremblay, H., Glangine, G.: Computing a lower bound for the solution of a Robotic Process Automation (RPA) problem using network flows. In: 2020 7th International Conference on Control, Decision and Information Technologies (CoDIT), pp. 118–123. IEEE, Prague (2020)
37. Guacales-Gualavisi, M., Salazar-Fierro, F., García-Santillán, J., Arciniega-Hidrobo, S., García-Santillán, I.: Computer system based on robotic process automation for detecting low student performance. In: International Conference on Information Technology and Systems. IEEE, Bogota (2020)
38. Krüger, M., Helmers, I.: Robotic process automation in der Energiewirtschaft. In: Doleski, O. (ed.) Realisierung Utility 4.0 Band 1, pp. 759–768. Springer, Wiesbaden (2020). https://doi.org/10.1007/978-3-658-25332-5_46
39. Wanner, J., Hofmann, A., Fischer, M., Imgrund, F., Janiesch, C., Geyer-Klingeberg, J.: Process selection in RPA projects – towards a quantifiable method of decision making. In: International Conference on Information Systems. AIS, Munich (2019)
40. Herm, L.-V., Janiesch, C., Reijers, H.A., Seubert, F.: From symbolic RPA to intelligent RPA: challenges for developing and operating intelligent software robots. In: 19th International Conference on Business Process Management. Springer, Rome (2021). ISBN 978-3-030-85468-3
41. Kohli, R., Grover, V.: Business value of IT: an essay on expanding research directions to keep up with the times. J. Assoc. Inf. Syst. **9**(1), 23–39 (2008)
42. Martins, P., Sa, F., Morgado, F., Cunha, C.: Using machine learning for cognitive Robotic Process Automation (RPA). In: 15th Iberian Conference on Information Systems and Technologies (CISTI), pp. 1–6. IEEE, Seville (2020)
43. Justus, D., Brennan, J., Bonner, S., McGough, A.S.: Predicting the computational cost of deep learning models. In: IEEE International Conference on Big Data (Big Data), pp. 3873–3882. IEEE (2018)

Adding Decision Management to Robotic Process Automation

Maximilian Völker(✉), Simon Siegert, and Mathias Weske

Hasso Plattner Institute, University of Potsdam, Potsdam, Germany
simon.siegert@student.hpi.de, {maximilian.voelker,mathias.weske}@hpi.de

Abstract. Robotic Process Automation promises to release employees from repetitive and monotonous work, providing space for creative and innovative tasks. RPA tools provide a wide range of techniques to automate user interactions, including filling forms and copying values between applications. While it is accepted that decisions play an important role in business processes, they are not a first-class citizen in RPA. This paper proposes a framework and a software architecture that integrates decision management into RPA. The work is evaluated by a prototype that introduces Decision Model and Notation (DMN) capabilities to the RPA software tool UiPath by utilizing Camunda's decision engine.

Keywords: Robotic Process Automation · Decision management · RPA design

1 Introduction

Over the last few years, Robotic Process Automation (RPA) gained momentum in research, fueled by its adoption in industry [1]. With the promise of taking digital, repetitive work off the hands of employees, companies are increasingly using RPA to improve the efficiency of their processes while freeing up human labor for more creative and innovative tasks [3,23]. Using software robots, RPA tools imitate the behavior of human users, such as mouse and keyboard inputs, but are also able to, for example, query web services and control applications [14,23]. As artificial intelligence techniques have improved, the scope of RPA applications has continued to expand. Nowadays, RPA can not only imitate interactions but also gains more and more human capabilities, such as text recognition and learning from past executions [7,14,19].

Despite the progress in terms of functionality, the bot development process did not change much. Mostly targeting low-code or no-code developers, many RPA tools offer graphical user interfaces to create new bot workflows using predefined building blocks [14].

In [22] we have observed that RPA vendors often only support *if/else* or *switch* constructs to steer the process. Thus, processes with complex decision logic are either very cumbersome to implement or the potential RPA process

J. González Enríquez et al. (Eds.): BPM 2021, LNBIP 428, pp. 23–37, 2021.
https://doi.org/10.1007/978-3-030-85867-4_3

may even be discarded for automation, which motivates the need for better decision management in RPA.

Similar issues were encountered for the business process modeling standard BPMN, such that decision-intensive processes led to complex process models with many nested branches [4, 24]. In traditional business process management, the Decision Model and Notation (DMN) standard [21] was introduced to resolve the problem by separating the decision logic from the process flow [4]. In this work, we examine whether and how elements of DMN, as a proven modeling standard for decisions, could be integrated into RPA. For this purpose, we analyze the RPA lifecycle and transfer elements from DMN to RPA to benefit from its powerful but still visual representation of decision logic.

After an introduction to DMN and RPA presented in Sect. 2, a motivating example is introduced in Sect. 3. For the integration, the RPA lifecycle is analyzed and related to DMN in Sect. 4, outlining potential synergies, and a generic software architecture is described. In Sect. 5, a prototype is presented that demonstrates the integration of DMN in the RPA vendor UiPath, enabling new use cases that were previously difficult to realize. Additionally, limitations of the approach are discussed. Section 6 summarizes the contribution and provides hints for future research.

2   Preliminaries

In this section, preliminary knowledge about the decision model and notation standard as well as robotic process automation is provided.

2.1   Decision Model and Notation

Modeling complex decisions using control flow often results in large, spaghetti-like models as it, for example, has been reported for the Business Process Model and Notation (BPMN) standard [24]. While such models are capable of correctly representing the decision logic, they are not only difficult to maintain, e.g., when a particular aspect in the decision logic changes, but their complexity also impedes communication using the model [4].

To solve this issue, the Decision Model and Notation (DMN) standard [21] was introduced. It allows to separate the decision logic from the control flow logic in process models represented in BPMN [20, 21]. DMN enables the automated evaluation of decisions and can therefore be used in automated business processes. Nevertheless, the standard also focuses on comprehensibility for non-technical users [10].

DMN provides various notation elements for representing highly complex decisions [21]. The central element is the *Decision Table*, which specifies the rules on how to derive the correct decision result from given input values. More specifically, a decision table consists of (i) a set of input parameters required for making the decision, (ii) a set of output variables whose values have to be determined based on the input parameters, as well as (iii) a list of rules that match values or value-ranges of the input parameters and assign the appropriate values to the output variables. So-called hit policies are used to specify how these rules are evaluated, e.g., whether only the first matching rule should be applied or the outputs of all matching rules should be returned.

An example for a decision table is given in the subsequent section in Fig. 3.

The DMN standard comprises additional elements, such as Decision Requirements Graphs, which are not further considered here.

2.2 Robotic Process Automation

While BPMN focuses on larger, often interdepartmental or even cross-organizational processes, the comparably new technology of Robotic Process Automation (RPA) concentrates on local workflows mainly on a single workstation [11]. The main goal of RPA is to automate frequent and rule-based workflows performed by a user on a computer [3,15,23], such as transferring data between different systems, like from an e-mail to a customer-relationship-management program. On the one hand, this is intended to relieve the user of such repetitive, monotonous tasks; on the other hand, it is expected to reduce the error rate and thus increase the overall quality [23]. To perform such automation, so-called RPA bots are utilized, small software clients that imitate the behavior of the user, e.g., by simulating mouse and keyboard interactions or more advanced operations [3,23].

In this context, many RPA software vendors target business users, i.e., automation with RPA should ideally be possible quickly and preferably without any programming knowledge [9,17]. Thus, many providers offer a graphical user interface to create RPA bots by combining predefined "building blocks" and thereby specifying a flow of individual automation operations [14].

The steps to introduce robotic process automation are reflected in the RPA lifecycle introduced by Jimenez-Ramirez et al. [16], which is given in Fig. 1.

The lifecycle enables the governance of entire RPA projects and ensures that the RPA software's performance is increased iteratively.

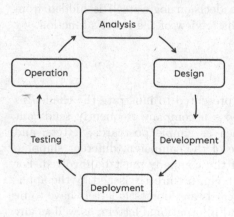

Fig. 1. RPA lifecycle (cf. [16])

The lifecycle starts with a context *analysis* phase to identify suitable processes for automation. Subsequently, the previously selected processes are further specified and modeled for automation in the *design* phase. In the *development* phase, these models are converted into executable programs, which are eventually run in the *deployment* phase. After the deployment, the bots are checked for errors in the *testing* phase, whereas in the subsequent *monitoring* phase, the robots are further operated and maintained. Gained performance metrics and insights into errors are then included in the next iteration of the lifecycle.

2.3 Decisions in RPA

For designing and developing RPA bots there is, unlike for business processes, no standard notation. Consequently, various vendor-specific syntaxes have emerged and therefore the support for different types of decision points in RPA bots differs from vendor to vendor. Table 1 shows the different possibilities for modeling decisions in selected, leading RPA tools, that are supported natively, i.e., are offered as building blocks for bots by default. All investigated providers offer basic *if/else* nodes with two outgoing control flow branches. Also, most of them provide the possibility to prompt the user a dialog with an input field or selection, which could be used to defer decisions to a human worker. More advanced constructs, such as *if/elseif* or *case* statements, i.e., elements with more than two outputs, are already less common and implemented with varying complexity.

Table 1. Native decision capabilities of some RPA providers

Decision Type	UiPath	Blue Prism	Automation Anywhere	Robot Framework
Human dialog	Yes	No	Yes	Yes
If/Else (2 outputs)	Yes	Yes	Yes	Yes
If/ElseIf/Else	No	No	Yes	Yes
Switch/Case	Yes	Yes	No	No

As a result, RPA encounters the same problem as BPMN. Workflows with more complex decisions lead to bloated models that are hard to understand and maintain [22], which is a problem because RPA is supposed to take over rule-based processes and be easy to use. Of course, similarly to script activities in BPMN, many RPA vendors offer blocks for executing custom code in which decision logic could be realized. While this might be feasible for certain use cases, in general, it contradicts the philosophy of RPA to be accessible even without programming knowledge. In addition, the decision logic itself is hidden from business users, who thus cannot gain a holistic view of the robot's function.

3 Motivating Example

In the following, a motivating example is presented to illustrate the challenges of decision-making in RPA models. Suppose a company frequently sends out advertising material in various forms, such as simple postcards, letters, and parcels. Depending on the form and scope of the campaign, different shipping costs accrue, and different departments of the company must be involved. For example, postcards for a national campaign can be directly issued by the applicant for 50 ct per piece, international postcards and domestic letters have to be commissioned via the sales department, and international letters, as well as any parcels, must be arranged with the logistics.

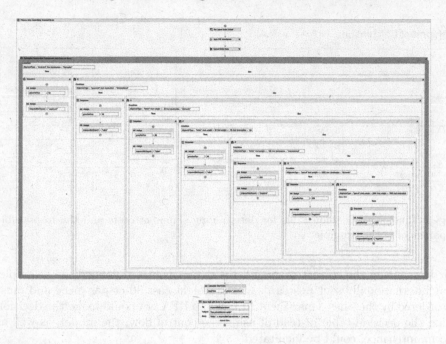

Fig. 2. RPA bot modeled with conventional decision elements (intentionally not readable) (Color figure online)

So far, the company has used a paper-based approach, i.e., the secretary received requests for a certain campaign, calculated the costs, and arranged the handover to the appropriate department. To simplify and streamline the communication process as well as to track requests, the company recently switched to digital documents, such that applications are now sent as PDF files. Still, the processing of the application remained a manual and humdrum task.

With RPA, automating the secretary's workflow described above becomes feasible, as RPA is able to read and send e-mails with attachments and analyze PDF files, especially if they are uniformly formatted, like forms.

However, the resulting RPA process is complex and lengthy as the decision logic to determine the price and the department has to be modeled using the above-described elements, like *if/else*, and the decision part, therefore, takes up a majority of the model as shown in Fig. 2 (blue-framed box). Once created, the bot is of course functional, however, its maintenance is difficult. For example, as soon as the production and shipment costs for the material change, the RPA bot needs to be updated, but the nested structure hampers quick adjustments, not to mention more fundamental changes in the decision logic. Consequently, it is very time-consuming to create and maintain the RPA bot, and it is very likely that the use case is soon discarded for automation with established RPA solutions.

Using DMN, however, the previously described decision can be modeled in a comprehensible and compact way, as shown in Fig. 3. Here, the individual rules

Shipment Calculation	Hit Policy: Unique ⌄			
When	And	And	Then	And
shipment type	weight	destination	price	responsibility
"parcel","letter","postcard"	integer	"domestic","international"	integer	"logistics","sales","applicant"
1 "postcard"	-	"domestic"	50	"applicant"
2 "postcard"	-	"international"	90	"sales"
3 "letter"	<= 30	"domestic"	80	"sales"
4 "letter"	]30..50]	"domestic"	95	"sales"
5 "letter"	<= 100	"international"	250	"logistics"
6 "parcel"	<= 2000	"domestic"	450	"logistics"
7 "parcel"	]2000..5000]	"domestic"	600	"logistics"

Fig. 3. Illustrative decision table for determining shipment costs and the responsible department in the example

for determining the correct costs and the department are defined, for example, that internationally sent postcards of any weight cost 80 ct per piece and must be ordered by the sales department. If now the RPA bot could make the decision using the decision table instead of using the control flow, the model, as well as the maintenance, could be facilitated.

4 Integration Concepts

In this section, the feasibility of integrating DMN, with focus on decision tables, within RPA is explored. For this, each phase of the RPA lifecycle, as introduced in Sect. 2.2, is examined for potential barriers and problems, and solution concepts are presented, with particular focus on the prominent phases of design and development as well as the execution time.

For the *analysis* phase, i.e., the selection of processes that are suitable for automation with RPA, different frameworks were proposed. In general, RPA is found especially useful for less complex processes to allow for short implementation times [23]. In terms of decision complexity, the integration of DMN can increase the number of suitable processes, since decisions that previously had to be laboriously modeled using control flow elements can then be created independently using DMN. Of course, not all decisions are suitable for automation with RPA and DMN. DMN is mainly suitable for modeling and making operational decisions that are well-defined, frequently executed, and rather have a local and short impact [6]. However, these requirements fit well with the characteristics of RPA processes, such as a high volume and degree of standardization, and the "digitized structured data input" [23], important for an automated evaluation of decisions.

4.1 Design

After the selection of a suitable process, the *design* phase involves creating a visual RPA process model that defines the RPA agent's relevant activities, structure, and data flow [16]. These models also enable communication and exchange about the behavior that the RPA bot should exhibit.

The overall goal of integrating DMN into RPA is to separate the decision logic from the control flow. This separation is therefore particularly apparent and significant in the design phase. The logic for decision-making, previously defined using the available bot building blocks such as if/else, should now be extracted into a single RPA activity dedicated to decision-making by evaluating a corresponding decision table.

In general, RPA bots are created in an RPA vendor-specific model notation that provides the activities available for automation and allows selecting and arranging them in a process-like sequence. But, as mentioned before, there is no standardized modeling language for RPA. Instead, each provider of RPA tools maintains its specific solution of a graphical or textual notation to represent the model. However, it is recommended to prefer intuitive visual modeling tools, as they do not require specific IT development skills and thus make the creation of RPA process models accessible to domain experts [12]. For the integration of DMN in RPA, graphical models are primarily suitable since DMN allows decisions to be represented by graphical elements and aims at being accessible to non-IT users as well [21].

For creating RPA models in the respective model notation, several ways are conceivable. Recent approaches suggest mining or learning RPA bots from past executions (i.e., [2, 13]). However, these approaches are currently at the beginning of their development and are rarely adopted in industry tools. Hence, they will not be considered further here.

Consequently, the traditional manual modeling of RPA bots is still the predominant way. It involves human workers, domain experts, as well as technical experts [23]. The integration of DMN benefits from this setting, as all stakeholders can participate in the modeling process, and thus all decision-relevant factors can be considered due to the broad circle of participants. Decision activities can easily be added to the RPA process by hand just as any other type of activity is added to the process.

Besides the manual modeling, most RPA tools offer a recording mode that tracks all interactions that a user performs on a computer system [18], which is a quick way to capture an RPA process [12]. This method, called screen recording, uses the observed user interactions to create a model that the RPA bot can then repeat exactly. However, screen recordings can only capture one execution path, i.e., a case-specific, linear workflow performed by a human user that does not include any choices regarding control flow or data. Therefore, this model creation method is suitable for repetitive processes without variation but not for processes with extensive decisions.

At the current state, the conventional modeling tends to be the most robust way to obtain an RPA process model with decision points. To enable better decision management for the screen recording approach as well, the recording functionalities would need to be adapted to support alternative execution branches.

In addition to the RPA model, the decision logic must now also be modeled, since, as already mentioned, the integration of DMN into RPA aims at separating the decision logic from the control flow. Here, the separation facilitates the communication regarding the decision logic, since it is not covered in the RPA bot model but explicitly represented using the DMN standard, enabling the collaboration of both business users and IT experts.

Similar to the RPA process, the modeling of the decision logic is mostly done manually. However, there are also approaches for mining decision logic from event logs [4,8]. While this could result in beneficial synergies with the previously mentioned mining methods of RPA models, it is still a complex challenge to extract both the process and decisions together from an event log [10] and requires more research.

Rather than testing robots only in their execution environment [16], the proposed integration enables initial verifications of RPA applications already at design-time using an existing formal property for decisions in business processes, decision soundness [5]. Based on decisions defined in DMN decision tables, asserting decision soundness increases the quality of RPA bots. As a result, run-time problems such as deadlocks will not occur. Decision soundness is based on the following criteria [5]:

- *Table Completeness:* any combination of inputs can be assigned to an output. E.g., the table in Fig. 3 is not complete as not every weight can be sent and the combination of *parcel* and *international* is not covered.
- *Output Coverage:* the process can handle all outputs of the decision. E.g., the bot can handle all possible values for the responsible department.
- *Dead Branch Absence:* any branch of the process flow after the decision point is reachable. E.g., check that there is no control flow branch in the bot for handling a value for the responsibility (like *promotion*), which was not defined in the table and could therefore never be reached.

While it is conceivable that these criteria could be checked in the context of RPA, it is, due to a lacking standard, heavily dependent on the chosen RPA vendor and requires further research. Nevertheless, such a formal verification at design-time could prevent avoidable errors during execution.

4.2 Development

The models created in the design phase are converted into executable code during the *development* phase. For the integration of DMN in RPA, additional requirements for the RPA tool's infrastructure must be met here, such as creating, storing, and evaluating decision tables.

DMN models are typically created within a separate modeling software [10]. To enable the handling of DMN decisions within RPA, either a DMN modeling tool needs to be developed, or an existing one needs to be accessed from the RPA software architecture.

As now separate models for the decisions are created, these models must be managed and stored next to the RPA bot models so that they are accessible from the RPA system architecture. A decision model repository provides this functionality. Within this, previously created decision models can be accessed, and ideally also versioned, allowing for updates and rollbacks. Also, a repository might be created as a central component for models which facilitates the reuse of decision tables within different RPA agents.

4.3   Deployment, Testing, and Operation

At run-time, not only the RPA bots must be enacted, but also decisions must be evaluated as soon as a bot reaches a decision point. Therefore, a so-called decision engine is required that can evaluate decision tables.

Evaluation means that for given input variables the corresponding output is calculated according to the rules in the table. For this purpose, the decision table must be parsed and processed at the software robot's run-time.

In general, for the evaluation of decision tables in RPA, two alternatives are conceivable, either a local embedding of a decision engine within the RPA tool or the connection to an external decision service. When the decision engine is directly embedded in the RPA software, no further, external dependencies are needed and decisions are created and evaluated locally. However, it requires the vendor to implement DMN functionalities that might already be present in potentially used BPMS systems.

Therefore, another approach is to outsource the modeling and evaluation of decisions to an external decision service. Such decision services usually already provide a modeling tool for defining decisions and a network interface through which third-party software can request their evaluation. To enable the processing of decision tables in RPA, the bot must only be able to connect to this interface to provide the required input values and obtain the decision's outputs. As the decisions are not defined within the RPA tool anymore, unlike the local embedding, this external approach facilitates sharing the same decision logic between RPA bots or even business processes of the company.

If the second approach is chosen, the external decision engine needs to be available beginning with the *deployment* phase of the RPA bot to ensure a connection can be established when required.

For the *testing* and *operation* phases, when the RPA bots are actually executed, the procedure for evaluating a decision, independent of an external or internal decision engine, is described in more detail in Sect. 4.4 (cf. Fig. 5).

In the *testing* phase, special attention should be paid to the decision points to ensure that required input data is available and also in the correct format so that the decisions can be evaluated as planned. During *operation*, where the performance metrics of the RPA bot are measured [16], additional performance indicators can now be derived, such as the distributions of decision outcomes, e.g., what share international letters account for in the example, which can be used to further improve the process in the subsequent passes of the lifecycle.

4.4   Generic Architecture

So far, two different approaches for integrating a DMN-based decision service
into RPA were discussed, either directly embedded in the RPA bot or by using
an external service.

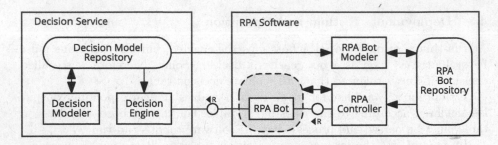

Fig. 4. Architecture with an external decision service

In Fig. 4, a conceivable, generic architecture for the external approach is
given. First, at design-time, the decision logic is defined using the *decision modeler* of the external decision service and subsequently saved in its *decision model
repository*. Now, whenever a new *RPA bot* is created in the RPA software, a
DMN decision point can be added to the bot's workflow, provided that such a
DMN activity is available in the RPA tool. This DMN activity is then linked to a
decision model stored in the decision model repository. Additionally, it requires
configuration of variables that should be passed from the bot to the decision service for evaluation, i.e., values available in the bot need to be mapped to input
values required by the decision.

At run-time, the two components communicate as shown in Fig. 5. As soon
as a bot, started and operated by the *RPA controller*, reaches a decision point,
it requests the *decision engine* of the external decision service using the linked
decision identifier and provides the required decision variables as input as configured at design-time. The decision engine then requests the decision table from
the decision model repository using the identifier and subsequently calculates
the decision result based on the input data and the decision table. The output
of the decision (*decisionResult*) is then returned to the bot so that the RPA
process can continue accordingly and use the decision result.

When using a locally embedded decision engine, a similar flow of communication would be applicable. With regard to the architecture, however, the RPA
software would need to be extended by the components required for providing DMN capabilities, i.e., a modeler (if not integrated in the bot modeler), an
engine for evaluating decisions, and a repository for storing the decision models,
to substitute the decision service.

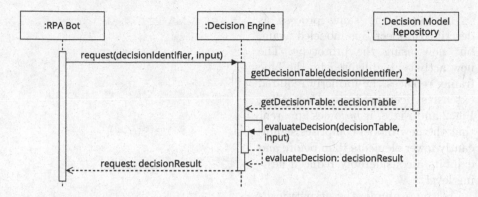

Fig. 5. Communication between RPA bot, decision engine, and decision model repository

5 Evaluation

The analysis of the lifecycle has shown that DMN and its decision tables, can, with some adjustments, be integrated into RPA. In this section, a potential realization of the integration, based on the generic architecture introduced in Sect. 4.4, is presented.

5.1 Proof of Concept Prototype

To demonstrate the feasibility of the proposed approach, we implemented[1] a DMN activity for the RPA tool UiPath[2] using the external decision service approach. For the decision service, Camunda[3] with its open-source modeler and decision engine is used.

The new DMN activity is available to the RPA bot creator as a normal building block and can be added to the workflow as usual, similar to the business rule task available in BPMN. When using the activity, the creator has to specify the internet address of the decision service and the identifier of the decision that should be evaluated (provided by the decision model repository). Furthermore, the variables of the RPA bot that should be passed to the decision engine at run-time must be provided, as well as the variables in which the output of the decision should be saved.

At execution-time, the activity requests the evaluation of the configured decision and supplies the current data stored in the variables to the Camunda engine. After evaluation, the result is interpreted and provided to the RPA bot in the specified variables, which can then be used in the subsequent flow.

---

[1] The prototype is open source and can be found on GitHub:
https://github.com/bptlab/rpa-dmn-operation.
[2] https://www.uipath.com/.
[3] https://camunda.com.

In Fig. 6, the same process as described in Sect. 3 is modeled again, but now using the prototype. The new activity, highlighted by the blue frame, replaces the formerly required extensive decision logic. Comparing Fig. 2 and Fig. 6, it becomes apparent that the new model comprises significantly fewer elements than before and exhibits a considerably reduced nesting level.

In the example, the activity calls the decision engine to evaluate the decision table given in Fig. 3. The output values of the decision, *pricePerPiece* and *responsibleDepartment*, are used subsequently to calculate the total costs and notify the appropriate departments. But the values could, for example, also be used to trigger different control flows depending on the *responsibleDepartment*, e.g., by using the *switch* statement.

By using the external decision service, i.e., a centralized solution, decisions can be reused in other bots as well, or existing decisions already used in business processes become available for use in RPA bots. Furthermore, this eases the maintenance of decisions, as the decision logic only has to be updated in one central place, instead of in all bots separately.

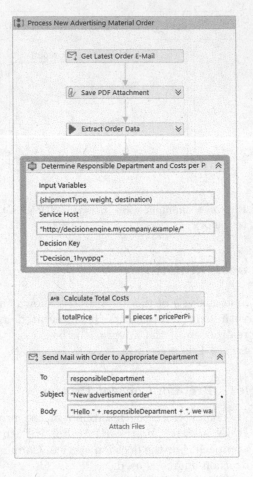

Fig. 6. Same RPA bot as shown in Fig. 2, but now modeled using the new DMN activity

5.2   Limitations

The presented prototype already enables RPA developers to separate decision logic and control flow. However, in the chosen approach, an external dependency is introduced. In this case, the decision logic is neither modeled nor evaluated within the RPA software, but relies on third-party software. This could be mitigated by directly integrating DMN capabilities into the RPA tool. This way, decisions could be modeled in the same tool as the RPA bot.

Overall, the integration of DMN into RPA not only increases the number of potential use cases, but also the complexity. Robotic process automation thrives on being easily accessible and quickly employed, without the need for extensive training. With DMN and its decision tables, another modeling standard must be mastered if this extension is to be used. However, especially for companies already employing BPMN, the DMN standard might already be familiar.

Furthermore, this approach is limited to data-based and rule-based decisions, as it inherits the limitations of DMN. Therefore, decisions that, for example, are of strategic nature or require human intuition, cannot be covered. Additionally, like in BPMN, the decision task itself does not branch the control flow, but, based on the decision result, the branching must still be modeled in the RPA bot using the available concepts. Nevertheless, encapsulating the decision-making process already reduces the complexity of the model to some extent.

6   Conclusion

Even though RPA promises to take over rule-based and routine tasks, the rudimentary support for making decisions in workflows may be an exclusion criterion for many decision-intensive processes. Future combinations with artificial intelligence are also expected to provide opportunities for improved and intelligent decision-making [7,23,25], such as learning decisions from past executions. However, it is unlikely that they will completely replace manual modeling and no longer require human intervention.

In this paper, we examined the integration of DMN, a standard for modeling and evaluating decisions, into RPA to address this limitation of current tools and analyzed the RPA lifecycle accordingly. Furthermore, an implementation for an RPA software was presented that allows bot creators to embed DMN decision points in bot workflows and subsequently use the decision result for further actions.

The integration of DMN in RPA offers several benefits. The size of bot models decreases as the decision logic does not have to be realized using control flow elements but is encapsulated in a decision task. This not only facilitates the modeling process itself, but also ensures better maintainability later, as the control flow logic and decision logic can be updated independently. In addition, especially if BPMN and DMN are already in use, it allows reusing decision logic in other bots or business processes, thus having a central place for decision logic. Overall, it may further increase the adoption of RPA, as the barriers for automating workflows with complex, data-based decision logic are lowered.

So far, the approach requires the decision tables to be crafted manually. However, the use of already existing approaches for mining decision logic from data could be evaluated further in the future. This would coincide with the recent endeavors to mine RPA bots from logs. Furthermore, we concentrated on decision tables in this work, but DMN provides more advanced concepts for decision management, such as decision requirements graphs, that could be considered in the future. Other interesting points for future research are checks for correctness or soundness of RPA bots in conjunction with DMN activities, as it has been done for BPMN and DMN. This is especially important since RPA bots are usually not tested in a separate environment, but are directly deployed to the live systems.

References

1. van der Aalst, W.M.P., Bichler, M., Heinzl, A.: Robotic Process Automation. Bus. Inf. Sys. Eng. **60**(4), 269–272 (2018). https://doi.org/10.1007/s12599-018-0542-4
2. Agostinelli, S., Lupia, M., Marrella, A., Mecella, M.: Automated generation of executable RPA scripts from user interface logs. In: Asatiani, A., et al. (eds.) BPM Blockchain and RPA Forum 2020. LNBIP, vol. 393, pp. 116–131. Springer, Cham (2020). https://doi.org/10.1007/978-3-030-58779-6_8
3. Aguirre, S., Rodriguez, A.: Automation of a business process using robotic process automation (RPA): a case study. In: Figueroa-García, J.C., López-Santana, E.R., Villa-Ramírez, J.L., Ferro-Escobar, R. (eds.) WEA 2017. CCIS, vol. 742, pp. 65–71. Springer, Cham (2017). https://doi.org/10.1007/978-3-319-66963-2_7
4. Batoulis, K., Meyer, A., Bazhenova, E., Decker, G., Weske, M.: Extracting decision logic from process models. In: Zdravkovic, J., Kirikova, M., Johannesson, P. (eds.) CAiSE 2015. LNCS, vol. 9097, pp. 349–366. Springer, Cham (2015). https://doi.org/10.1007/978-3-319-19069-3_22
5. Batoulis, K., Weske, M.: Soundness of decision-aware business processes. In: Carmona, J., Engels, G., Kumar, A. (eds.) BPM Forum 2017. LNBIP, vol. 297, pp. 106–124. Springer, Cham (2017). https://doi.org/10.1007/978-3-319-65015-9_7
6. Biard, T., Le Mauff, A., Bigand, M., Bourey, J.-P.: Separation of decision modeling from business process modeling using new "decision model and notation" (DMN) for automating operational decision-making. In: Camarinha-Matos, L.M., Bénaben, F., Picard, W. (eds.) PRO-VE 2015. IAICT, vol. 463, pp. 489–496. Springer, Cham (2015). https://doi.org/10.1007/978-3-319-24141-8_45
7. Chakraborti, T., et al.: From robotic process automation to intelligent process automation. In: Asatiani, A., et al. (eds.) BPM Blockchain and RPA Forum 2020. LNBIP, vol. 393, pp. 215–228. Springer, Cham (2020). https://doi.org/10.1007/978-3-030-58779-6_15
8. De Smedt, J., Hasić, F., vanden Broucke, S.K.L.M., Vanthienen, J.: Towards a holistic discovery of decisions in process-aware information systems. In: Carmona, J., Engels, G., Kumar, A. (eds.) BPM 2017. LNCS, vol. 10445, pp. 183–199. Springer, Cham (2017). https://doi.org/10.1007/978-3-319-65000-5_11
9. Enriquez, J.G., Jimenez-Ramirez, A., Dominguez-Mayo, F.J., Garcia-Garcia, J.A.: Robotic process automation: a scientific and industrial systematic mapping study. IEEE Access **8**, 39113–39129 (2020)
10. Figl, K., Mendling, J., Tokdemir, G., Vanthienen, J.: What we know and what we do not know about DMN. EMISAJ **13**(2), 1–16 (2018)
11. Flechsig, C., Lohmer, J., Lasch, R.: Realizing the full potential of robotic process automation through a combination with BPM. In: Bierwirth, C., Kirschstein, T., Sackmann, D. (eds.) Logistics Management. LNL, pp. 104–119. Springer, Cham (2019). https://doi.org/10.1007/978-3-030-29821-0_8
12. Galusha, B.: Considering RPA? Ask smart questions for long-term success. Database Trends Appl. **31**, 44–45 (2017)
13. Gao, J., van Zelst, S.J., Lu, X., van der Aalst, W.M.P.: Automated robotic process automation: a self-learning approach. In: Panetto, H., Debruyne, C., Hepp, M., Lewis, D., Ardagna, C.A., Meersman, R. (eds.) OTM 2019. LNCS, vol. 11877, pp. 95–112. Springer, Cham (2019). https://doi.org/10.1007/978-3-030-33246-4_6
14. Hofmann, P., Samp, C., Urbach, N.: Robotic process automation. Electron. Mark. **30**(1), 99–106 (2019). https://doi.org/10.1007/s12525-019-00365-8

15. Ivančić, L., Suša Vugec, D., Bosilj Vukšić, V.: Robotic process automation: systematic literature review. In: Di Ciccio, C., Staples, M., et al. (eds.) BPM 2019 Blockchain and CEE Forum. LNBIP, vol. 361, pp. 280–295. Springer, Cham (2019). https://doi.org/10.1007/978-3-030-30429-4_19
16. Jimenez-Ramirez, A., Reijers, H.A., Barba, I., Del Valle, C.: A method to improve the early stages of the robotic process automation lifecycle. In: Giorgini, P., Weber, B. (eds.) CAiSE 2019. LNCS, vol. 11483, pp. 446–461. Springer, Cham (2019). https://doi.org/10.1007/978-3-030-21290-2_28
17. Lacity, M.C., Willcocks, L.P.: A new approach to automating services. MIT Sloan Manage. Rev. Fall **58**, 41–49 (2017)
18. Leno, V.: Multi-perspective process model discovery for robotic process automation. In: CAiSE 2018 Doctoral Consortium. CEUR-WS.org (2018)
19. Martínez-Rojas, A., Barba, I., Enríquez, J.G.: Towards a taxonomy of cognitive RPA components. In: Asatiani, A., García, J.M., Helander, N., Jiménez-Ramírez, A., Koschmider, A., Mendling, J., Meroni, G., Reijers, H.A. (eds.) BPM Blockchain and RPA Forum 2020. LNBIP, vol. 393, pp. 161–175. Springer, Cham (2020). https://doi.org/10.1007/978-3-030-58779-6_11
20. Object Management Group: Business Process Model and Notation (BPMN) (2014). https://www.omg.org/spec/BPMN/
21. Object Management Group: Decision Model and Notation (DMN) (2021). https://www.omg.org/spec/DMN
22. Siegert, S., Völker, M.: Towards decision management for robotic process automation. In: ZEUS 2021, pp. 9–13. CEUR-WS.org (2021)
23. Syed, R., et al.: Robotic process automation: contemporary themes and challenges. Comput. Ind. **115**, 103162 (2020)
24. van der Aa, H., Leopold, H., Batoulis, K., Weske, M., Reijers, H.A.: Integrated process and decision modeling for data-driven processes. In: Reichert, M., Reijers, H.A. (eds.) BPM Workshops 2015. LNBIP, vol. 256, pp. 405–417. Springer, Cham (2016). https://doi.org/10.1007/978-3-319-42887-1_33
25. Viehhauser, J.: Is robotic process automation becoming intelligent? Early evidence of influences of artificial intelligence on robotic process automation. In: Asatiani, A., et al. (eds.) BPM 2020. LNBIP, vol. 393, pp. 101–115. Springer, Cham (2020). https://doi.org/10.1007/978-3-030-58779-6_7

AIRPA: An Architecture to Support the Execution and Maintenance of AI-Powered RPA Robots

A. Martínez-Rojas[1]([⊠]), J. Sánchez-Oliva[2], J. M. López-Carnicer[1],
and A. Jiménez-Ramírez[1]

[1] Departamento de Lenguajes y Sistemas Informáticos, Escuela Técnica Superior de
Ingeniería Informática, Avenida Reina Mercedes, s/n, 41012 Sevilla, Spain
{amrojas,ajramirez}@us.es, jose.lopez@iwt2.org
[2] Servinform, S.A. Parque Industrial PISA, Calle Manufactura,
5, Mairena del Aljarafe, 41927 Sevilla, Spain
jmsanchezo@servinform.es

Abstract. Robotic Process Automation (RPA) has quickly evolved
from automating simple rule-based tasks. Nowadays, RPA is required
to mimic more sophisticated human tasks, thus implying its combination with Artificial Intelligence (AI) technology, i.e., the so-called intelligent RPA. Putting together RPA with AI leads to a challenging scenario
since (1) it involves professionals from both fields who typically have
different skills and backgrounds, and (2) AI models tend to degrade
over time which affects the performance of the overall solution. This
paper describes the AIRPA project, which addresses these challenges by
proposing a software architecture that enables (1) the abstraction of the
robot development from the AI development and (2) the monitor, control, and maintain intelligent RPA developments to ensure its quality
and performance over time. The project has been conducted in the Servinform context, a Spanish consultancy firm, and the proposed prototype
has been validated with reality settings. The initial experiences yield
promising results in reducing AHT (Average Handle Time) in processes
where AIRPA deployed cognitive robots, which encourages exploring the
support of intelligent RPA development.

Keywords: Robotic Process Automation · Artificial Intelligence ·
Industrial project

1 Introduction

The term Robotic Process Automation (RPA) refers to a software paradigm in
which robots are programs that mimic the behavior of human workers interacting

This research has been supported by the NICO project (PID2019-105455GB-C31) of
the Spanish Ministry of Science, Innovation and Universities and the AIRPA project
(EXP00118029/IDI-20190524, P011-19/E09) of the Center for the Development of
Industrial Technology (CDTI).

J. González Enríquez et al. (Eds.): BPM 2021, LNBIP 428, pp. 38–48, 2021.
https://doi.org/10.1007/978-3-030-85867-4_4

with information systems (IS) [3,13]. This paradigm has become increasingly popular because RPA is of great interest to organizations [5].

In this context, RPA solutions based on Artificial Intelligence (AI) – called intelligent RPA solutions – are receiving increasing attention, as the combination of both disciplines offers and several advantages [2]. On the one hand, AI methods enhance RPA solutions by providing new capabilities that enable a more significant number of end-to-end processes to be automated. On the other hand, RPA solutions produce data on the execution of the process themselves, allowing periodic training of the AI models, leading to continuous improvement of the model metrics [9].

The use of this kind of component involves different challenges when a methodology, architecture o role specification proposal does not exist.

First, a data scientist (i.e., a professional in charge of processing structured data to extract relevant information from it) is required to develop the cognitive components and, without an abstraction role, needs to know the business process to configure a model for each business case. This fact leads to a strong dependency between the data scientist and the RPA developer role (i.e., professional in charge of designing and developing software robots) who must know the business process to automate it[1]. Secondly, the performance of these components in a production environment depends on the data model performance, which tends to degrade over time [6]. Degradation refers to multiple reasons, such as: (1) the evolution of the business over time and the AI obsolescence in the new business context, (2) the AI technology advance caused by the new scientist research in the AI, that can improve the AI performance and accuracy, led previous AI models obsolete, and (3) the need to re-training models to increase their accuracy and performance for a specific task. This problem arises the need to conduct the AIRPA project, a platform with an architecture that allows solving the challenges encountered, (1) separate the work of RPA developer and data scientist to abstract robot construction from model development, and (2) control, monitor, and support the robots to ensure quality maintenance of the intelligent RPA components.

As shown in Fig. 1, RPA developers are in charge of building the robot that automates the process. However, such developers may lack skills related to AI. For this, AI services (e.g., text-to-image, speech-to-text recognition, sentiment analysis, image anomaly detection, and others) have to be black boxes that always maintain an acceptable level of accuracy in their responses. The data scientist's implementation of these cognitive solutions makes their use transparent to the RPA developer. To these needs, the AIRPA project provides an architecture that supports the abstraction between both roles, and continuous monitoring mechanisms to ensure the quality of AI models, for both new release deployments and retraining.

The rest of the paper is organized as follows. Section 2 describes the project context and a set of example cases. Sect. 3 presents the AIRPA project. Section 4

---

[1] https://www.edureka.co/blog/rpa-developer-roles-and-responsibilities/.

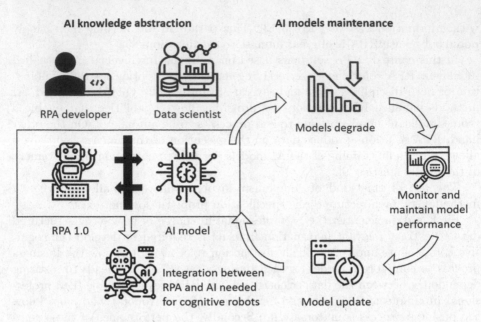

Fig. 1. Main challenges to be solved when making use of cognitive components.

briefly summarizes related work. Section 5 opens a discussion related to the project. Finally, Sect. 6 concludes the paper and describes future work.

2   Context

The RPA challenges described in Fig. 1 have also been pointed out by the industry, expressly, by Servinform S.A.[2] Servinform is a Spanish company dedicated to providing outsourcing services to other companies, mainly back-office processes automated with RPA. In the past years, they have identified the need to implement intelligent RPA solutions to empower business processes.

Integrating cognitive components in their processes allows automating tasks that previously required human intervention, i.e., aiming towards end-to-end process automation. For instance, the prediction of electricity consumption, considering that electricity use changes over time and consumption forecasts are of great value to utilities. For this purpose, a component is needed that determines what the consumption will be on the next bill, based on a customer's data history. Another example is related to document classification since companies typically use different formats when they issue documents. In this way, a component is required for classifying invoices or sales orders from different organizations with similar information but a different structure or style.

---

[2] https://www.servinform.es/.

The use of these components within an RPA process presents a series of challenges (cf. Fig. 1). As shown in the second challenge, a problem related to the AI components degradation and the reduction of ML model performance has been found. That is, this paper will focus its proposed architecture on the fact that the *"maintenance of a machine learning model involves regular updating to ensure that predictive effectiveness is not lost over time"* [6]. Therefore, Servinform, together with the IWT2 research group[3], tries to solve these challenges with the AIRPA research project, which will be described below.

3 Research Project

This section explains the AIRPA project. First, the initial objectives are presented. Second, the approach proposed in the project is detailed. Finally, the architecture to be developed and put into production is described.

3.1 Initial Goals

Based on Servinform industrial experience and the background within the RPA research line of the IWT2 group, the following goals are identified as the main ones of the AIRPA project:[4]

1. Create a collection of AI components to empower RPA solutions.
2. Create a nexus of union between both domains by understandably presenting the results in a platform for technical staff and business experts.
3. Automatize processes end-to-end that facilitate the integration between existing RPA solutions with AI components, reducing the need for human participation and decision-making.
4. Simplify and reduce the cost of access to RPA solutions powered with AI caused by licensing restrictions.
5. Enable RPA professionals who lack AI and ML skills to use AI components.
6. Define a lifecycle, development methodology, production, and integration roadmap of RPA solutions with AI components.
7. Verify the developed AIRPA framework in multiples realistic scenarios.
8. Integrate an AI components library in RPA solutions. Such integration seeks sustainability based on a cross-platform architecture orchestration independent of specific technologies and considers the degradation of AI over time.

3.2 Approach

The AIRPA project proposes a complete solution for the implementation of RPA processes using cognitive components, known in the industry as intelligent RPA processes.[5] To this end, it defines an architecture that supports its development

---

[3] https://www.iwt2.org/.

[4] As the project is under development, the realization of their goals are in progress.

[5] https://dlabs.ai/blog/rpa-2-0-how-to-achieve-the-highest-level-of-automation/.

and maintenance, divided into four modules (cf. Fig. 2): (1) *Document repository*, where components and robots are stored with their documentation and all their associated versions, (2) *Deployment manager*, which is used to control the deployment and version management of each component and the RPA robots, (3) *Tracking and exploitation panel*, which allows the visualization of the metrics and data associated with the execution of the processes, especially for the monitoring of the models that are associated with the cognitive components, and (4) *Control Room*, which allows for complete management of RPA processes, and handling cases in an execution state, KOs (i.e., failed situations), robots, equipment where they are executed, customized alerts, evidence capture, launches, or user roles, among others.

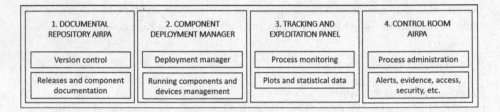

Fig. 2. Modules that compose the AIRPA platform.

The *control room* aims to provide comprehensive and centralized support to the intelligent RPA process management. This module resembles a customized state machine for each process, indicating which state a robot is in at any given moment and recording evidence of its transitions. The essential tasks of this module are (1) the collection of data needed for evidence capture and (2) the management and automated reporting of robots and tasks in which human intervention is required. The latter provides a differential value in intelligent RPA processes, thanks to labeling the data collected in daily work. In other words, the management of hybrid human-robot tasks allows the capture of data from the decisions which are made by humans, an essential task for the training of cognitive component models that will fully automate these tasks in the future. The collection of this data, together with the information reported by the cognitive components, feeds the *tracking and exploitation panel* for process reporting and obtaining valuable information of each process during its execution. It monitors, among other things, the performance of the models, facilitating their maintenance.

The first challenge focuses on the abstraction of the complexity of the development of an AI component by the RPA developer. For this purpose, *AIRPA components* are built, i.e., AI services with a microservices-based architecture that standardizes their service contract through a REST API. These components are designed as *wrappers* that allow the incorporation of ML models using the files previously exported in the data scientist's work environment. In this way, the RPA developer should only focus on consuming the methods offered by this API. Thus, any changes to the model will not affect its integration with the RPA solution.

Additionally, AI components designed as *wrappers* solve further problems. Generally, AI components in the context of intelligent RPA are implemented by commercial solutions (e.g., Amazon Web Services, Google, or Microsoft). These solutions pose several problems since a series of compatibility restrictions limit their use. They are not versatile enough to re-train the models from business data, and their customization becomes a rather complicated task. The *wrappers* component design allows the incorporation of proprietary Machine Learning (ML) models, solving these issues. This fact provides an added value since, although their use is not widespread in the context of RPA, open-source solutions are leading the main developments in the field of AI [12]. This design increases the specialization capacity of each model and reduces the cost of access caused by licensing limitations to AI components. Therefore, the *AIRPA components* of the AIRPA project, a library representing the first initial objective of the project, enable ML solutions such as classifications, anomaly detection, intelligent document processing, audio transcription, or sentiment analysis, among others.

3.3   Architecture

The AIRPA project proposes an architecture (cf. Fig. 3) for the execution of RPA processes that use cognitive components and that allows uploading, deploying, managing, and monitoring both robots and AI components. This architecture has different types of developments.

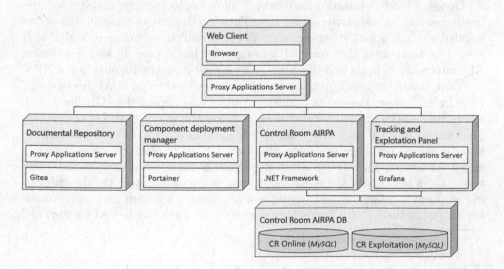

Fig. 3. Architecture diagram.

Firstly, the modules are based on several free software solutions as *Gitea* for the Document repository, *Portainer* for the Deployment manager, and *Grafana* for the Tracking and exploitation panel. Secondly, some solutions are based on customized development, as the *Control Room* implemented in *C#* with the *.NET* framework and the *MySQL* database. This module has two databases, the *CR Online* for the management of all information necessary to use *Control Room* (e.g., users, alerts, scheduled events, etc.) and the *CR Exploitation* with the data collection of all activity of the deployed robots and AI components to analyze them. Another customized solution is the web client developed with *ASP.NET Web Pages (Razor)* and offers access to all the modules that comprise the AIRPA architecture from a unique site that permits login to the whole system. Finally, the creation of *wrappers* for the *AIRPA Components* is a custom design implemented in *Python* language with *Django REST Framework*. In its construction, several specific libraries from the field of data processing and ML are used to facilitate feature engineering and the incorporation of ML models from *Scikit Learn*, *TensorFlow*, and *PyTorch* based on a service contract.

3.4 Achievements of Goals

The current status of the AIRPA platform shows the degree of accomplishment of the initial objectives of the project.

4 Related Work

In the current industry, some platforms aim to solve problems related to the one addressed in this paper. However, they are oriented to different perspectives.

Google Cloud[6], Amazon[7] and Azure[8] allow deploying and monitoring pre-created or custom cognitive components. These platforms can monitor and detect cognitive degradation. However, they are not unified to be used in RPA. It should be noted that they offer RPA integration but deploying and monitoring AI components is in an isolation system separated from the monitoring of RPA.

That makes the monitoring more complex due to the use of AI models, e.g., Blue Prism[9] allows the use of cognitive components deployed in Google Cloud, where they are monitored. Nonetheless, it forces the use of both (1) the Google platform to monitor cognitive components and (2) the Blue Prism platform to monitoring the non-cognitive ones. Similarly, UIPath[10] platform offers a service called *AI Center*. The service allows deploying cognitive components and monitoring them but, unlike the AIRPA platform, it does not detect the degradation of components. Moreover, *AI Center* is a proprietary solution and only accessible for use in the UIPath technology stack (i.e., UIPath Studio, AI Center, and

---

[6]   https://cloud.google.com/vertex-ai.
[7]   https://aws.amazon.com/sagemaker/.
[8]   https://azure.microsoft.com/en-gb/services/machine-learning/.
[9]   https://www.blueprism.com/.
[10]   https://www.uipath.com/.

Table 1. Table of the completed initials goals.

Goal	State	Comments
1	Completed	An AI components library described in Sect. 3.2 is developed
2	Completed	Thanks to the *tracking and exploitation panel*, graphs and statistics can be presented that are easy to understand for business experts
3	Completed	The use of the *control room* gives full support to process automation and, therefore, to end-to-end automation, which reduces the number of people needed to solve process cases and reduces costs
4	Completed	The proposed design enables to use open-source solutions, which actually lead the field of AI, increasing the specialization capacity of each model and reducing the cost of access caused by licensing limitation
5	Completed	The proposed architecture abstracts RPA developer from AI models development and maintenance, through the use of *wrappers*
6	In validation	A robotization guide is being defined to be applied for the design, development and maintenance of each component, still pending to be validated
7	In progress	The AIRPA project is still in the validation phase, but has been tested with some real scenarios as shown in Sect. 5
8	Completed	AIRPA architecture allows the integration of AI components in RPA processes from controlling the status of the robots to monitoring their performance.

Orchestrator). In other words, the existing RPA platforms do not support the detection of model degradation in AI components. So it is necessary to navigate to AI service platforms providers to consult this information. Moreover, since AI services are not integrated with RPA platforms, it is necessary to manage logins, tokens, etc. between the two platforms. Furthermore, AIRPA differentiates itself from other platforms by the use of AI components as *wrappers*. With these, APIs through AI components are offered, they follow the same specification and, therefore, RPA developers only need to know this specification to implement the AI-RPA integration (Table 1).

Besides these commercial solutions, there are scientific research proposals that study the RPA and the AI field [4,13]. Some of them use AI not to build components, but for early stages in the RPA lifecycle, in the process discovery phases [3,7,10,11]. Nevertheless, others deal with the application of AI on RPA processes, such as [1], which studies key open research challenges that exist in the combination of RPA with AI, or [9], which proposes a dynamic taxonomy for intelligent RPA components. There are other proposals to improve the RPA architecture. For example, [8] elaborates on optimizing the deployment architecture of the RPA components. Nevertheless, to the best of our knowledge, there are no proposals that focus on the operation and maintenance phase of the robots, such as AIRPA. At a glance, unlike existing proposals, AIRPA proposes a system to manage and intensively monitor cognitive components, separating the AI development from the RPA developer. All these are carried out easily and integrally in one platform.

5   Results

The AIRPA project is still in the validation phase, and therefore the final results may vary after it. The platform has been validated in several real scenarios that require the incorporation of cognitive actions. In this way, the aim is to evaluate the AIRPA operation applied to different business areas, such as energy or telecommunications. One of these validation has been carried out specifically on the consumption prediction use case shown in Sect. 2. In this case, the generation of automatic predictions from historical data, without the need for human interaction, represents a considerable improvement. It is performed within Servinform's operation area by taking measurements during one month of the Average Handle Time (AHT) before and after implementing the AIRPA platform. Initially, an AHT of 9 min was obtained, which was improved after implementing AIRPA by 75%, resulting in a final AHT of 2 min and 15 s. Even though the results are preliminary, the platform significantly increased control over the process, which suggests promising results. AIRPA is planned to be a platform and a methodological strategy followed by Servinform S.A. and its entire Consulting and Innovation area.

The conduction of the AIRPA project has lead to some lessons learned. After defining and implementing AIRPA architecture, we have realized that it is very focused on high-demand situations in terms of deployments. In some real scenarios, the level of demand for deployments is lower because they are less frequent. Therefore, in that cases, such a complex architecture is not necessary, and a deployment pipeline would be sufficient. In addition, the experience with different real cases showed limited use of the shared collection of RPA components. This situation was mainly since each case required a level of customization that neglects its transversality. Nonetheless, the *wrapper* design of the components enables easy customization from one business case to another.

6   Conclusions and Future Work

This paper presented the AIRPA project that aims to improve RPA thanks to the integration of AI, expanding the automation of the end-to-end processes. The project allows (1) the abstraction of the data scientist from the RPA developer and (2) extensive monitoring of robots and AI components to detect the degradation of the cognitive components. To solve these challenges, AIRPA has been built with a microservice architecture and the standardization of a service contract thanks to an API REST. These components are designed as *wrappers* to facilitate their integration. This architecture is composed of 4 modules: the document repository, the deployment manager, the AIRPA Control Room, and the tracking and exploitation panel. All these modules allow the RPA developers and the data expert to work independently. In addition, thanks to the Control Room and the tracking panel, the detection of AI degradation is possible.

Although the project is still in progress, it offers promising preliminary results and possible lines of future research work. (1) The project, being research-based, could have performance improvements and better functionalities. (2) The component library can be extended by adding new functionalities. (3) Currently, the data scientist requires an existing *wrapper* component before loading a model. The need to avoid the dependency of the data scientist from the RPA developer role and ease the modification of components is identified, so the use of preloaded *wrappers* is proposed as future work. (4) The current *Control Room* is a custom state machine (cf. Sect. 3.2) and the robot is the one in charge of modifying the state. As future work, we plan to extend the component behavior to automatically generate multiple state changes and evidence transitions records.

References

1. Chakraborti, T., Isahagian, V., Khalaf, R., Khazaeni, Y., Muthusamy, V., Rizk, Y., Unuvar, M.: From robotic process automation to intelligent process automation: emerging trends. In: Internation Conference on Business Process Management 2020 RPA Forum. Springer, Cham (2020). https://doi.org/10.1007/978-3-030-58779-6
2. Jha, N., Prashar, D., Nagpal, A.: Combining artificial intelligence with robotic process automation—an intelligent automation approach. In: Ahmed, K.R., Hassanien, A.E. (eds.) Deep Learning and Big Data for Intelligent Transportation. SCI, vol. 945, pp. 245–264. Springer, Cham (2021). https://doi.org/10.1007/978-3-030-65661-4_12
3. Jimenez-Ramirez, A., Reijers, H.A., Barba, I., Del Valle, C.: A method to improve the early stages of the robotic process automation lifecycle. In: Giorgini, P., Weber, B. (eds.) CAiSE 2019. LNCS, vol. 11483, pp. 446–461. Springer, Cham (2019). https://doi.org/10.1007/978-3-030-21290-2_28
4. Hollebeek, L.D., Sprott, D.E., Brady,M.K.: Rise of the machines? Customer engagement in automated service interactions. J. Serv. Res. 24(1), 3–8 (2021). https://doi.org/10.1177/1094670520975110, https://www.sciencedirect.com/science/article/pii/0166361596000139
5. Le Clair, C., O'Donnell, G., Lipson, A., Lynch, D.: The Forrester Wave™: Robotic Process Automation, Q4 2019. The Forrester Wave (2019)
6. Leevy, J.L., Khoshgoftaar, T.M., Bauder, R.A., Seliya, N.: The effect of time on the maintenance of a predictive model. In: 2019 18th IEEE International Conference On Machine Learning And Applications (ICMLA), pp. 1891–1896. IEEE (2019)
7. Leno, V., Polyvyanyy, A., Dumas, M., La Rosa, M., Maggi, F.M.: Robotic process mining vision and challenges. Bus. Inf. Syst. Eng. 63,301–314 (2020)
8. Mahala, G., Sindhgatta, R., Dam, H.K., Ghose, A.: Designing optimal robotic process automation architectures. In: Kafeza, E., Benatallah, B., Martinelli, F., Hacid, H., Bouguettaya, A., Motahari, H. (eds.) ICSOC 2020. LNCS, vol. 12571, pp. 448–456. Springer, Cham (2020). https://doi.org/10.1007/978-3-030-65310-1_32

9. Martínez-Rojas, A., Barba, I., Enríquez, J.G.: Towards a taxonomy of cognitive RPA components. In: Asatiani, A., García, J.M., Helander, N., Jiménez-Ramírez, A., Koschmider, A., Mendling, J., Meroni, G., Reijers, H.A. (eds.) BPM 2020. LNBIP, vol. 393, pp. 161–175. Springer, Cham (2020). https://doi.org/10.1007/978-3-030-58779-6_11

10. Martins, P., Sá, F., Morgado, F., Cunha, C.: Using machine learning for cognitive Robotic Process Automation (RPA). In: Iberian Conference on Information Systems and Technologies (CISTI) (2020)

11. Singh, M.K., Raghavendra, D., Pandian, D., Sadana, A.: Surface automation - interacting with applications using Black box approach. In: International Conference for Convergence in Technology (I2CT) (2021)

12. Sonnenburg, S., et al.: The need for open source software in machine learning. J. Mach. Learn. Res. **8**,2443–2466 (2007)

13. Willcocks, L., Lacity, M., Craig, A.: Robotic process automation: strategic transformation lever for global business services? J. Inf. Technol. Teach. Cases **7**(1), 17–28 (2017)

Blockchain Forum

An Empirical Evaluation of Smart Contract-Based Data Quality Assessment in Ethereum

Marco Comuzzi[1]([✉]), Cinzia Cappiello[2], and Giovanni Meroni[2]

[1] Ulsan National Institute of Science and Technology, Ulsan, Republic of Korea
mcomuzzi@unist.ac.kr
[2] Politecnico di Milano, Milan, Italy
{cinzia.cappiello,giovanni.meroni}@polimi.it

Abstract. The data carried by transaction payloads play a crucial role in smart contract-based blockchain systems. Therefore, blockchains should be equipped with mechanisms to control their data quality. In practice, however, such mechanisms are currently missing. While in our previous work we have proposed how data quality controls can be implemented as smart contracts, in this paper we focus specifically on the evaluation of their execution overhead (time and cost). Evaluating this overhead is crucial to understand in which situations the cost of controlling the data quality of transaction payloads can be sustained by a blockchain system. We have implemented in Ethereum two pseudo-real scenarios that cover all the types of data quality controls in blockchains that we defined in our previous work and evaluated for each of them the time and cost overhead. The results show that the overhead of control can be high particularly for controls involving oracles that fetch off-chain data and controls that require to correlate data from different transactions.

Keywords: Blockchain · Data quality · Cost · Smart contract · Ethereum

1 Introduction

Smart contract-enabled blockchains increasingly underpin the implementation of resilient and trustless distributed information systems. Examples of such systems are supply chain management platforms, open data registries, and electricity trading and billing platforms in smart grids [12]. In these systems, the data carried by transactions payloads determine which data are stored in the distributed ledger and which application logic (smart contracts) is executed by all nodes of the network.

Given the crucial role of transaction payloads, we would expect smart contract-enabled blockchains to be equipped with mechanisms that guarantee the data quality of these payloads, i.e., their fitness for use [6]. For example, in a cold supply chain scenario, a sensor that reports a temperature reading greater

J. González Enríquez et al. (Eds.): BPM 2021, LNBIP 428, pp. 51–66, 2021.
https://doi.org/10.1007/978-3-030-85867-4_5

than 25% of the previous recorded value or that falls outside a range of admissible temperatures is transmitting an inaccurate value. Such an inaccuracy may indicate a fault in the sensor or a problem with the transportation process. In both cases, data quality assessment would highlight the anomalous value and trigger further analysis before it can be accepted. Therefore, mechanisms to implement data quality controls and to eventually discard low quality data should be implemented. So far, blockchains provide natively only some primitive mechanisms to guarantee such quality: in cryptocurrencies, transactions are validated by nodes receiving them only to check if users own the coins that are transferring [2].

As the quality of the data heavily influences the reliability of the applications that use them, data quality controls performed on-chain can significantly increase the users' trust in blockchain applications. In our previous work [7], we have proposed an approach to implement data quality controls on-chain using ad-hoc smart contracts in Ethereum. Specifically, the approach implements data quality smart contract templates addressing different quality aspects (i.e., dimensions) combining (i) the type of data required by data quality controls, e.g., whether a control requires a single value of a variable only or multiple time series of multiple variables, and (ii) the way in which these data are delivered to the blockchain, e.g., whether by one or multiple transactions.

A crucial concern when implementing data quality assessment for blockchains is to evaluate the *overhead* of its execution. Blockchains, particularly public ones, can have in fact a high cost and time overhead. For instance, users have to pay to use a public blockchain, such as in the form of transaction fees collected by miner nodes in systems that use proof-of-work consensus. The evaluation of the overhead of data quality assessment remains an open issue. In the case of ad-hoc smart contracts for data quality control considered in this paper, the overhead of data quality assessment can be a combination of the monetary cost of the fees required to execute the data quality assessment smart contracts, e.g., the amount of gas required to run them in Ethereum, and the additional time that may be required to execute the data quality control, which impacts the time for a transaction to be mined into a block and, therefore, the system throughput.

In this context, the contributions of this paper are: (i) To present in detail two pseudo-real scenarios of ad-hoc smart contracts for data quality assessment, which cover all the type of smart contract templates for data quality assessment identified by our previous research [7]; (ii) To empirically evaluate the overhead (cost and time) of implementing data quality assessment using the two identified pseudo-real scenarios in Ethereum. The results obtained show that the overhead of control can be high particularly for controls involving oracles that fetch off-chain data and controls that require to correlate data from different transactions.

The paper is organised as follows. Section 2 briefly surveys the related work in data quality on blockchain. Section 2 discusses the related work. Section 3 introduces the data quality model, summarising also our previous research. The scenarios and smart contracts considered in the evaluation are presented in detail in Sect. 4, while the results of the evaluation are discussed in Sect. 5. Conclusions are finally drawn in Sect. 6.

2 Background and Related Work

Data quality (DQ) is often defined as the capability of data to satisfy the users' requirements [6]. Being a multidimensional concept, DQ is evaluated taking different DQ *dimensions* into consideration. The most commonly used dimensions are *accuracy* – the degree with which data values are correct – *completeness* – measuring the degree with which required data are present in a dataset – *timeliness* – measuring the temporal validity of data – and *consistency* – measuring the degree with which data are valid according to defined rules, such as functional dependencies or business rules. The *metrics* used to evaluate DQ dimensions may vary depending on the type of data and data source. For instance, assessing the accuracy of strings requires a different metric than assessing the accuracy of numbers.

DQ assessments can be performed either *online* or *offline*. Online DQ assessments occur when new data are saved into a storage system. The objective in this case is to deal immediately with low-quality data, for instance, by rejecting them to avoid lowering the overall quality of the data in the storage system. Conversely, offline DQ assessments operate after the data have been stored, either periodically or when the storage system is queried. This paper focuses on the online assessment of the DQ quality of transaction payloads.

Second-generation blockchains, such as Ethereum and Hyperledger Fabric, support the so-called *smart contracts* [10], executable code capturing, in a broad sense, how business is to be conducted among organizations, e.g., the transfer of digital assets after a condition is fulfilled. Nodes can invoke smart contracts by issuing transactions that specify the operation of the contract to be invoked and optional parameters in their payload. As far as incentives are concerned, when a new transaction is processed, the node that issued it is billed proportionally to the amount of data contained in the transaction and to the complexity of the invoked smart contract operation.

The research on the quality of blockchain applications focuses on *software quality*. For example, Atzei et al. [3] classify code vulnerabilities in Ethereum smart contracts. Wohrer and Zdun [11] outline security patterns for smart contracts. Bartoletti and Pompianu [5] identify common programming patterns in Ethereum smart contracts based on the type of application.

Given their ability to store data in a consistent, immutable and persistent form across multiple nodes, blockchains often are seen as systems that can effectively improve data quality [9,13]. However, blockchain research often assumes that the data stored in the blockchain are correct. While this assumption holds for data that are created from the blockchain itself (i.e., cryptocurrency), the quality of data created outside the blockchain may vary, and our work aims to guarantee that only reliable data are shared within a blockchain system. Thus, in the literature, the issue of data quality in blockchain research has not been explored in depth yet. Chen et al. [9] argue that, in most application scenarios, the use of a blockchain alone may already increase data integrity and quality. This is supported by Azaria et al. [4], who discuss the implementation of a medical record management system using blockchain, observing an improvement

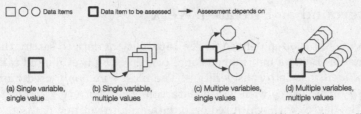

Dependencies among variables and values when assessing data quality: single variable, single value vs. multiple variables and time series of values

Fig. 1. Dependencies among variables and values when assessing DQ. Reprinted from [7].

of both quality and quantity of the data stored for medical research. Casado-Vara et al. [8] study data quality in blockchains for IoT applications. Despite the research cited above, a generic, application-independent approach to control data quality, which specifically takes into account the strengths and limitations of a blockchain, is still missing.

3   Data Quality Assessment Model

This section introduces a model of DQ controls in blockchains (Sect. 3.1) and then summarises our previous work on smart contract-based data quality assessment (Sect. 3.2).

3.1   Data Quality Controls in Blockchains

As discussed in Sect. 2, DQ can be assessed using different DQ dimensions and each dimension can be associated with multiple assessment metrics. In a given context, the DQ assessment logic depends on the type of sources and on the type of data, and it may require additional metadata (e.g., expected values, consistency rules). Considering such information needs, in our previous work we defined four situations that may occur (see Fig. 1):

- Single variable, single value (SS): the quality assessment of a variable does not require additional data;
- Single variable, multiple values (SM): the quality evaluation of a value depends on the availability of one or more historical values of the same variable;
- Multiple variables, single value per variable (MS): the quality of a value relies on single values of a number of other variables;
- Multiple variables, multiple values per variable (MM): the quality of a value relies on multiple values of a number of other variables.

These situations influence the way in which we model transactions and DQ controls, as described in the following.

As regards transactions, we consider proof-of-work blockchain systems that use a client-generated nonce to order transactions, such as Ethereum. Transactions are submitted by client applications to one node of a blockchain network. They carry data items, i.e., key-value pairs. Therefore, a transaction $t = \langle n, [d_i], c \rangle$ is defined by a nonce n, the set of data items d_i that it carries, and a correlation id c. Each d_i is the key associate to a value v_i. The nonce n is an incremental value specified by the issuer of the transaction, which is used to order transactions. A transaction is considered valid only if its nonce has not already been used by the same issuer, otherwise it is discarded. Also, a transaction is processed only after all the other transactions from the same issuer with nonce lower than the current one have been received. The correlation id c is required to match data items referring to the same instance of stateful data quality controls when multiple values are needed.

A *quality control* is defined as $dqc = \langle logic, [input_j], action \rangle$, where *logic* contains the assessment logic of the considered DQ dimensions, $[input_j]$ is the list of input parameters of the assessment logic, i.e., a list of data items and *action* defines the task to perform in case of poor quality. Input parameters are received as data items d_i in transactions. Data quality controls can be *stateless* or *stateful*. In a stateless *dqc*, the data items $input_j$ are carried by a single transactions. Therefore, *dqc* can run as soon as this transaction is received. In a stateful *dqc*, the data items $input_j$ are carried by different transactions. Therefore, *dqc* can run only once all the transactions carried the required $input_j$ have been received.

Specifically, a new instance of a stateful quality control dqc_c is created when the first data item $d_i \in input_j$ is carried by a transaction t with correlation id c. The correlation id c is required to match data items from different transactions referring to the same instance of a quality control. For instance, a *dqc* may check the precision of three consecutive pressure readings from a sensor, in which case a correlation id can be generated combining the sensor id and a reading's timestamp[1]. A *dqc* can be evaluated once a data item for each input parameter in transaction carrying the same correlation id c has been received. The evaluation can be either positive or negative. We say that a data item d_i targets a *dqc* when it is required for its evaluation. For example, the quality control requiring a temperature value to be in a certain range is stateless, as it requires only data contained in the current transaction, while the aforementioned quality control requiring a temperature value not to be greater than 25% of the previous value is stateful, as it would require both the current transaction and the previous one.

The smart contract-based DQ assessment that we consider in this paper is not heavily affected by the application *logic* of DQ controls. The *action* to perform when poor quality is detected is to reject the transactions carrying low quality data (more nuanced policies to define actions are discussed in [7]).

---

[1] In principle, we could assume that each data item in a transaction can be associated with a different correlation id. For simplicity in this paper we consider that all data items carried by a transaction are associated with the same correlation id.

3.2   Smart Contract-Based Data Quality Assessment

To make the paper self-contained, this section gives a brief overview of the approach described in [7]. In a nutshell, in smart contract-based data quality assessment, the assessment of the DQ of transaction payloads is delegated to smart
contracts created ad-hoc to run DQ controls. The *functional* smart contract, i.e.,
the one invoked by a transaction carrying a data item $input_j$ required by a DQ
control dqc, is responsible for invoking the DQ smart contract implementing dqc
before executing any other functional logic that uses the data item $input_j$ as
parameters.

In this approach, the logic of DQ controls is implemented into functions
of a smart contract. Since smart contracts can be invoked by all nodes of a
blockchain, this solution enables every node to assess the quality of transaction
payloads based on their specific needs, e.g., the DQ requirements of the client
applications using them. DQ control functions are either stateless or stateful,
depending on the type of DQ control that they implement. Stateful ones exploit
the correlation id when necessary and require memory to store the values received
by different transactions. Both types of functions can use oracles to fetch off-
chain data required for the evaluation of DQ controls.

Our previous work [7] identifies a set of templates to support the implementation of data quality smart contracts, which serve four possible ways in which
the data items required by a DQ control (see Fig. 2) can be delivered to the
blockchain:

- Single transaction (ST): all the data required by a DQ control are contained
  in the payload of one individual transaction;
- Ordered transactions (OT): the data required by a DQ control are contained
  in different transactions that are received by all nodes in the same order. This
  situation occurs when the transactions are sufficiently spaced in time, such
  that it is possible to assume[2] that all nodes will receive them in the same
  order in which they are originated;
- Interleaved transactions (IT): the data required by a DQ control are contained
  in different transactions that may not be received in the same order by all
  nodes. In this case, we assume that the transaction payloads also contain
  information required by a DQ smart contract to understand when all the
  data required have been received, such as a counter or a correlation id;
- Off-chain (Off): the data required by a DQ control are available off-chain and
  injected into the blockchain via an oracle.

Considering the dependencies described in Fig. 1, it is clear that the case of
single variable, single value (SS) can only be associated with the single transaction (ST) scenario. Ordered transactions and interleaved transactions will be
used when the assessment logic is based on multiple values, while the off-chain
data could be used to retrieve single values of multiple variables.

---

[2] Note that this is not an absolute guarantee, because of the best-effort nature of the
Internet.

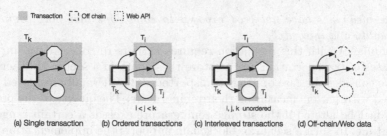

Fig. 2. Availability and correlation of data for quality assessment (using the multiple variables/single values configuration for presentation purpose; other configurations are similar). Reprinted from [7].

4   Two Scenarios for Data Quality Assessment

The two scenarios that we created for the DQ assessment overhead evaluation are discussed next. For each scenario, we have created a set of DQ control smart contract variants. These variants are classified according to the dimensions of the model presented in the previous section (Table 1 shows how the different combinations defined in the model are covered by the smart contract variants).

Table 1. Distribution of DQ smart contract variants over the dimensions of the DQ model. Depending on the DQ dimensions assessed, one smart contract may belong to different categories.

	(SS) Single var. Single values	(SM) Single var. Mult. values	(MS) Mult. var. Single values	(MM) Mult. var. Mult. values
(ST) Single transactions	SC1-2c, SC1-2d	SC1-1b	SC1-2a, SC1-2b	SC1-1a
(OT) Ordered transactions	N.A.	SC1-2c	SC1-2e SC2a, SC2b	SC1-1c SC1-2a, SC1-2e SC2a, SC2c
(IT) Interleaved transactions	N.A.	SC1-2d	SC1-2f	SC1-2b, SC1-2f
(Off) Off-chain data	N.A.	N.A.	SC2c, SC2d	SC2b, SC2d

4.1   Scenario 1: Drugs Transportation

In the EU, the transportation of medicinal products for human use is regulated by the GDP Regulation C343/01, which states that *"it is the responsibility of the supplying wholesale distributor to protect medicinal products against breakage, adulteration and theft, and to ensure that temperature conditions are maintained within acceptable limits during transport...it should be possible to demonstrate that the medicines have not been exposed to conditions that may compromise their quality and integrity".*

Compliance with this regulation requires carriers to collect data during all the phases of transportation and to store these data in a secure and persistent way. We consider the case of drugs transported using thermally-insulated trucks equipped with sensors measuring their temperature. The quality of the products may in fact be altered if the storage temperature exceeds a certain range for a certain time. In such a scenario, blockchain supports the implementation of the monitoring system: if stored on a blockchain, the data collected from the trucks and warehouses can not be lost or tampered with, helping carriers to avoid liability for not meeting the required transportation standards or for hiding employee negligence.

In this scenario, the assessment of the quality of sensor temperature readings concerns the following DQ dimensions: accuracy, timeliness, completeness, and precision. A boolean value is associated with each sensor reading for each dimension, indicating whether or not that reading satisfies the DQ requirements for that dimension. The DQ controls implemented for each dimension require additional DQ parameters (metadata), such as the range of admissible values for accuracy, the maximum delay acceptable for a temperature reading for timeliness, and a precision range for the precision DQ dimension. Note that the assessment of the accuracy and the timeliness require only the current sensor reading, whereas completeness and precision require also a set of previous sensor readings (the number of previous sensor readings to consider is an additional DQ parameter).

The 11 variants of the DQ assessment smart contracts for this scenario are summarised below.

SC1-1: Each transaction contains a batch of sensor readings and their timestamps. This is the first baseline scenario in which DQ assessment is not performed.

SC1-1a [(MM,ST)]: Each transaction contains a batch of sensor readings (MM), their timestamps, and the DQ parameters required to run the DQ controls. In this case, all the DQ controls find all the data items required inside one transaction (ST). All DQ dimensions are therefore immediately assessed when this transaction is mined into a new block.

SC1-1b [(SM,ST)]: Each transaction contains a batch of sensor readings and their timestamps (SM), while the DQ parameters required to run the DQ controls are immutably set when the smart contract is instantiated. In this case, all the DQ controls find all the required data items inside a transaction (ST). All the DQ dimensions are immediately assessed when the transaction is mined into a new block.

SC1-1c [(MM,OT)]: Each transaction contains a batch of sensor readings and their timestamps, while the DQ parameters required to run the DQ controls are sent in another transaction of a different kind, i.e., invoking another function of the DQ smart contract (MM). Therefore, all the DQ controls need to analyse both the current transaction containing sensor readings, and the last transaction received with values of the parameters. We assume that this transaction is received by all nodes before the one containing the batch of readings to be assessed (OT). All the DQ dimensions are immediately assessed when the transaction is mined in a new block.

SC1-2: Each transaction contains a single sensor reading and its timestamp. This is a second baseline scenario in which DQ assessment is not performed.

SC1-2a [(MS,ST),(MM,OT)]: Each transaction contains a single sensor reading, its timestamp, and the DQ parameters required to run the DQ controls (MS). In this case, accuracy and timeliness DQ controls find all the data items required in a transaction (ST). Conversely, completeness and precision DQ controls need to analyse also the previous transactions, which are sufficiently spaced in time to be received by all nodes in the same order (OT). All the DQ dimensions are immediately assessed when the transaction is mined in a new block.

SC1-2b [(MS,ST),(MM,IT)]: Each transaction contains a single sensor reading, its timestamp, a nonce (which is incremented every time a new transaction is submitted), and the DQ parameters required for the controls to operate. With respect to SC1b, this case does not take for granted that transactions may be mined in the same order as they were sent, and relies on the nonce to order the transactions (IT). Therefore, if a transaction is mined before the previous one, completeness and precision DQ controls are delayed until the previous transaction is mined (MM, IT). For accuracy and timeliness, DQ controls can be run as soon as a transaction is received (MS,ST).

SC1-2c [(SS,ST),(SM,OT)]: Each transaction contains a single sensor reading and its timestamp, while the DQ parameters required to run the DQ controls are immutably set when the smart contract is instantiated. In this case, the accuracy and timeliness DQ controls find all the data items that they require in a transaction (SS,ST). Conversely, the completeness and precision DQ controls need to analyse also the previous transactions, which we assume are mined in the correct order (SM,OT). All the DQ dimensions are immediately assessed when a transaction is mined into a new block.

SC1-2d [(SS,ST),(SM,IT)]: Each transaction contains a single sensor reading, its timestamp, and a nonce, while the DQ parameters required to run the DQ controls are immutably set when the smart contract is instantiated. As far as accuracy and timeliness are concerned, this variant is equivalent to SC1-2c (SS,ST). Regarding precision and completeness, and with respect to SC1e, this variant does not take for granted that transactions may be mined in the same order as they were sent (IT), and relies on the nonce to order the transactions. Therefore, if a transaction is mined before the previous one is, completeness and precision controls are delayed until the previous transaction is mined.

SC1-2e [(MS,OT),(MM,OT)]: Each transaction contains a single sensor reading and its timestamp, while the DQ parameters required for the DQ controls to run are sent in another transaction of a different kind. With respect to SC1-2d, the completeness and precision DQ controls also need to analyse the transactions containing the previous sensor readings. The data items required by the accuracy and timeliness DQ controls are all contained in a single transaction.

We assume that transactions are received in the same order by all nodes (OT). All DQ dimensions are immediately assessed when the transaction is mined in a new block.

SC1-2f [(MS,IT),(MM,IT)]: Each transaction contains a single sensor reading, its timestamp, and a nonce, whereas the DQ parameters required to run the DQ controls are sent, together with a nonce and a timestamp, in another transaction of a different kind. This is the most complex variant, since it needs to correlate transactions of different kind (sensor readings and DQ parameters) that may not arrive in the right order. To this aim, the nonce is used to sort the transactions, whereas the timestamp of the sensor readings and the one of the parameters act as a correlation identifier. It is worth noting that DQ dimensions can be immediately assessed only if all the transactions are mined in the correct order, and only for the sensor readings that were performed before the last DQ parameters update. In all the other cases, DQ controls are delayed.

4.2 Scenario 2: Drugs Prescription

General practitioners (GP) prescribe dozens of medications every day to their patients. Patients have their clinical background and (possibly) a list of ongoing treatments, so it can be difficult for GPs to assess whether a new prescribed drug can cause them any harm, for instance because it is incompatible with an ongoing treatment.

We assume that prescriptions are recorded in a blockchain system invoking a function of a smart contract. Each new prescription contains the patient's SSN and a prescribed drug. The DQ control functions of the smart contract, in this case, can be invoked manually by a GP for each prescription to check (i) if the patient SSN exists (Eligibility), and (ii) to verify whether there are any drugs already prescribed that are incompatible with the current prescription (Compatibility).

The 5 variants of the DQ control smart contracts for this scenario are summarized next:

SC2: This the baseline scenario in which DQ assessment is not performed.

SC2a [(MS,OT)(MM,OT)]: Both drug incompatibilities and eligible patients are stored on-chain. More in detail, each time a new patient is registered, a transaction containing the patient SSN is submitted. Similarly, when a new drug is introduced, another type of transaction containing the drug ID, its category and its incompatibilities is submitted. When a new prescription is created, another type of transaction containing the patient SSN and the drug being prescribed is submitted. Since all the information required by the DQ controls is available in the blockchain when this transaction is submitted, DQ controls are immediately performed once the transaction is mined.

SC2b [(MS,OT),(MM,Off)]: Eligible patients are stored on-chain, whereas drug incompatibilities are stored off-chain. With respect to SC2a, only transactions related with new patients and new prescriptions are submitted. Instead, the list of drugs, their categories and their incompatibilities are stored in an XML file published on an Internet-facing Web server. Therefore, whenever a new prescription is created, the smart contract must invoke an oracle that reads the XML file, retrieves the information related to the drug being prescribed, and notify this information to the smart contract. As a consequence, the eligibility control is immediately performed once a transaction is mined. Conversely, the compatibility control is on hold until the oracle calls back the smart contract providing to it the list of incompatibilities.

SC2c [(MS,Off),(MM,OT)] Eligible patients are stored off-chain, whereas drug incompatibilities are stored on-chain. With respect to SC2a, only transactions related to new drugs and new prescriptions are submitted. Instead, eligible patients are stored in an XML file published on an Internet-facing Web server. Therefore, whenever a new prescription is created, the smart contract must invoke an oracle that reads the XML file, finds if the patient is present in that file, and notifies that to the smart contract. As a consequence, the compatibility control is immediately performed once the transaction is mined. Conversely, the eligibility control is on hold until the oracle calls back the smart contract notifying it whether the patient exists or not.

SC2d [(MS,Off),(MM,Off)]: Both drug incompatibilities and eligible patients are stored off-chain. With respect to SC2a, only transactions related to new prescriptions are submitted. Instead, eligible patients and the list of drugs, their categories and their incompatibilities are stored in XML files published on an Internet-facing Web server. Therefore, whenever a new prescription is created, the smart contract must invoke an oracle that reads the XML files, finds if the patient is present in that file, retrieves the information related to the drug being prescribed, and notifies that to the smart contract. As a consequence, all DQ controls are on hold until the oracle calls back the smart contract notifying it if the patient exists and, if so, providing to it the list of incompatibilities.

5   Evaluation of Data Quality Assessment Overhead

We created a set of smart contracts implementing the DQ assessment variants discussed in the previous section in the Solidity language using the Remix IDE. The smart contracts have been deployed on the Ethereum Ropsten test network, which is an Ethereum public network where the Ether cryptocurrency is virtual, making smart contract invocations free of charge.

For Scenario 1, we simulated the behaviour of a sensor that performs a reading every 6 min. In particular, for SC1-1 and its variants, a transaction is submitted every 30 min, so a batch of 5 sensor readings is expected per transaction. Conversely, for SC1-2 and its variants, a transaction is submitted each time a sensor reading is performed. In addition, for SC1-2f, a transaction containing the updated DQ parameters is submitted every 30 min in order to trigger the DQ controls for the sensor readings received before that transaction. Finally, for SC1-2b, SC1-2d, and SC1-2f, we intentionally delayed the submission of 1 transaction out of 5 for 12 min, in order to test the ability of the smart contracts to deal with interleaved transactions.

For Scenario 2, we consider 8 drugs that could be prescribed to 3 eligible patients. For SC2a and SC2b, before the drugs could be prescribed, a transaction is submitted for each patient (to notify to the smart contract that the patients are eligible). Additionally, for SC2a and SC2c, before the drugs could be prescribed, a transaction is submitted for each drug (to notify to the smart contract which drugs are incompatible with each other). Conversely, for SC2b, SC2c, and SC2d, we also relied on oracles to access off-chain data for checking the existence of a drug (SC2b and SC2d) and for retrieving the category to which the drug belongs and the drugs with which it is incompatible (SC2c and SC2d). These oracles have been implemented using Provable (provable.xyz), an online service offering a library of functions that programmers can invoke to read data from an external source and push them inside smart contracts.

To evaluate the overhead of DQ smart contracts, we considered the following metrics:

- **Set-up gas.** The amount of gas needed to make the smart contract operational, which is a measure of both the complexity of the smart contract and the cost to deploy it. This metric, in fact, includes both the gas spent to deploy the smart contract, and the sum of the gas spent for invoking the functions that pass all the data required by the smart contract to perform the DQ assessments. For example, in SC1g the latter includes the gas spent to invoke the functions passing the DQ parameters, such as the expected temperature to compute the accuracy, whereas in SC2a it includes the gas spent to invoke the functions that store in the smart contract the eligible patients and the incompatibilities among drugs.
- **Processing gas.** The amount of gas needed on average to store the data collected into a smart contract, which is a measure of both the complexity of the smart contract and the cost it requires to be executed. For example, in SC1g, it comprises the gas spent to submit a new batch of sensor readings. In SC2a, it comprises the gas spent to submit a new prescription.
- **Processing time**. The amount of time one has to wait before the data collected are stored in a smart contract. For this metric, both the arithmetic mean and the standard deviation are computed.
- **DQ validation delay**. The amount of time one has to wait, once the data collected are stored in the smart contract, before their quality is assessed. For this metric, the arithmetic mean, the standard deviation, the absolute maximum and the absolute minimum are computed.

The code of the smart contracts is available at https://bitbucket.org/polimiisgroup/dq-solidity. To minimise the effects of congestion in the Ropsten test network, which causes significant variations in the time required for a transaction to be mined, we have repeated our measurements invoking the same smart contract functions in 7 different days, 5 times during business days and 2 times during the weekend, when the traffic is lower according to the data that we collected. The performance that we report is the average of the performance values registered in these 7 executions.

Table 2. Results of the experiments (time reported as mm:ss).

Smart contract	Set-up gas		Processing gas		Processing time			DQ validation delay			
	Value	Increment	Value	Increment	AVG	Increment	STDEV	AVG	STDEV	MIN	MAX
SC1-1	482785	N/A	324958	N/A	01:00	N/A	01:16	N/A	N/A	N/A	N/A
SC1-1a	1016659	+111%	333817	+3%	00:31	−48%	00:15	00:00	00:00	00:00	00:00
SC1-1b	1184398	+145%	346425	+7%	00:29	−52%	00:27	00:00	00:00	00:00	00:00
SC1-1c	1243773	+158%	346425	+7%	00:35	−42%	00:22	00:00	00:00	00:00	00:00
SC1-2	401441	N/A	104387	N/A	00:17	N/A	00:08	N/A	N/A	N/A	N/A
SC1-2a	838387	+109%	135331	+30%	00:45	+165%	00:43	00:00	00:00	00:00	00:00
SC1-2b	923172	+130%	140929	+35%	00:22	+29%	00:15	00:37	01:54	00:00	06:35
SC1-2c	1011508	+152%	147850	+42%	00:59	+247%	03:43	00:00	00:00	00:00	00:00
SC1-2d	1092466	+172%	152611	+46%	00:20	+18%	00:10	00:37	01:52	00:00	06:27
SC1-2e	1070942	+167%	147850	+42%	00:23	+35%	00:15	00:00	00:00	00:00	00:00
SC1-2f	1526459	+280%	132418	+27%	00:21	+24%	00:11	18:02	08:38	06:48	30:40
SC2	2748971	N/A	184566	N/A	00:22	N/A	00:25	N/A	N/A	N/A	N/A
SC2a	4342274	+58%	507400	+175%	01:17	+250%	01:37	00:00	00:00	00:00	00:00
SC2b	5251579	+91%	685506	+271%	00:24	+9%	00:19	00:26	00:17	00:01	00:56
SC2c	6112655	+122%	270056	+46%	00:28	+27%	00:29	00:06	00:13	00:00	00:40
SC2d	5249786	+91%	813988	+341%	00:36	+64%	00:26	00:33	00:14	00:12	00:57

Table 2 shows the results of the experiment, and allows the comparison of each smart contract with its variants that implement the DQ controls discussed in Sect. 4.

The results regarding the processing time have to be carefully interpreted. In practice, we have seen that, while the processing time is highly variable depending on the network conditions (see the standard deviation values in Table 2, which often are higher than the corresponding average), the impact of DQ controls on the processing time is negligible. That is, the processing time with and without DQ controls is normally comparable, and it follows the fluctuations of transaction processing time in the test network.

Concerning Scenario 1, the impact of DQ controls on the set-up gas is quite high, as it causes an increment ranging from 111% to 158% for SC1-1, and from 109% to 280% for SC1-2. Conversely, the impact of DQ controls on the processing gas is rather modest for SC1-1, as the increment ranges from 3% to 7%, while for SC1-2 is more substantial, as it ranges from 27% to 46%. It is worth noting that the DQ validation delay occurs only for SC1-2b, SC1-2d, and SC1-2f, as those are the only smart contracts that delay DQ controls in order to take into account interleaved transactions. In particular, for SC1-2b and SC1-2d the maximum delay corresponds to the difference between the time when the delayed transaction is mined and the time when the subsequent transaction is mined. Conversely, the minimum delay is 0 for transactions that are submitted before the time when the delayed one should have been submitted, and for those that are submitted after the delayed one was submitted. For SC1-2f, the maximum delay corresponds to the difference between the time when the first transaction after the last DQ parameters update is mined, and the time when the first transaction after the previous DQ parameters update is mined. Instead, the minimum delay corresponds to the difference between the time when the first transaction after the last DQ parameters update is mined, and the time when the last transaction before the last DQ parameters update is mined. Therefore, in SC1-2f, the DQ validation delay depends on the frequency of the DQ parameters updates.

Concerning Scenario 2, the impact of DQ controls on both the set-up gas and the processing gas is quite high, as it causes an increment ranging from 58% to 122% for the former, and from 46% to 341% for the latter. It is worth noting that the DQ validation delay occurs only for SC2b, SC2c, and SC2d, as they require external data to be fetched by Provable. Also, the DQ validation time equates on average to the processing time. This can be explained by the way Provable is invoked. Indeed, when contacted, Provable retrieves the off-chain data and submits them to the contract by submitting a transaction that invokes a callback function of the smart contract. Therefore, for DQ controls to take place, two subsequent transactions must be submitted, one with the data to evaluate, and another with the off-chain metadata for the DQ validation algorithm.

In summary, the results show that adding data quality controls to a smart contract has a considerable impact on both the gas used and the time-related performance to deploy and invoke it. The overhead in terms of gas used, in particular, is strictly related with the complexity of the code of the DQ control functions. Therefore, this overhead can be minimised by applying smart contract code optimisation techniques, e.g. [1], which are not specific to DQ controls. The overhead increases dramatically in an unpredictable way if oracles are used to fetch off-chain data required by a DQ control. However, it is worth noting that, although the variants of SC2 that make use of Provable are the most expensive in terms of gas, the amount of patients and drugs in this experiment is rather small. In a real scenario, where the contract may have to deal with thousands of patients and drugs, it could become inconvenient or even impossible not to rely on off-chain data and oracles.

6   Conclusions

We have presented an empirical evaluation of smart contract-based DQ assessment of transaction payloads in Ethereum. For the evaluation, we have developed a set of variants of DQ assessment smart contracts covering different types of DQ controls and measured the overhead associated with their execution on the Ethereum Ropsten test network.

The obtained results show that if DQ controls do not rely on oracles, the time required to validate transactions is low. Conversely, the impact in terms of resources needed to execute the DQ controls can be quite high, especially when the smart contracts are deployed in a public blockchain that requires cryptocurrency to operate. DQ controls that rely on oracles have a high impact on the amount of required resources. In our experiments, they required double the time in respect of the situation without oracles. Indeed, accessing oracles requires to submit two subsequent transactions: one related to the smart contract invocation and one related to the oracle invocation.

It is also worth noting that, the more DQ controls are resilient to network problems, such as interleaved transactions, the more expensive the smart contract will be and the more time it will take for DQ controls to be performed. Consequently, there is a trade-off between the accuracy of DQ controls and the time and cost required for such controls to operate.

We plan to extend this work in several ways. More scenarios and experiments can be implemented to evaluate the overhead more thoroughly, studying its variations for instance with the type of data quality control considered. More in general, existing data quality management methodologies can be extended to the case of blockchain as the core technology for implementing the application logic and/or storing data. The approach based on smart contracts can be tested also in other blockchains to check to what extent the results presented here can be generalized.

Acknowledgements. The authors thank Lorenzo Maria Bonelli for his help with the initial implementation of the smart contracts.

References

1. Albert, E., Correas, J. , Gordillo, P., Román-Díez, G., Rubio, A.: GASOL: gas analysis and optimization for ethereum smart contracts. In TACAS, pp.118–125. Springer (2020)
2. Antonopoulos, A.M., Wood, G.: Mastering Ethereum: Building Smart Contracts and Dapps. O'reilly Media, Gavin Wood (2018)
3. Atzei, N., Bartoletti, M., Cimoli, T.: A survey of attacks on ethereum smart contracts (SoK). In: Maffei, M., Ryan, M. (eds.) POST 2017. LNCS, vol. 10204, pp. 164–186. Springer, Heidelberg (2017). https://doi.org/10.1007/978-3-662-54455-6_8
4. Azaria, A., Ekblaw, A., Vieira, T., Lippman, A.: MedRec: using blockchain for medical data access and permission management. OBD **2016**, 25–30 (2016)
5. Bartoletti, M., Pompianu, L.: An empirical analysis of smart contracts: platforms, applications, and design patterns. In: Brenner, M., et al. (eds.) FC 2017. LNCS, vol. 10323, pp. 494–509. Springer, Cham (2017). https://doi.org/10.1007/978-3-319-70278-0_31
6. Batini, C., Scannapieco, M.: Data and Information Quality - Dimensions Principles and Techniques. Data-Centric Systems and Applications. Springer, Cham (2016). https://doi.org/10.1007/978-3-319-24106-7
7. Cappiello, C., Comuzzi, M., Daniel, F., Meroni, G.: Data quality control in blockchain applications. In: Di Ciccio, C., et al. (eds.) BPM 2019. LNBIP, vol. 361, pp. 166–181. Springer, Cham (2019). https://doi.org/10.1007/978-3-030-30429-4_12
8. Casado-Vara, R., de la Prieta, F., Prieto, J., Corchado, J.M.: Blockchain framework for IoT data quality via edge computing. In: BlockSys@SenSys 2018, pp. 19–24. ACM (2018)
9. Chen, S., Shi, R., Ren, Z., Yan, J., Shi, Y., Zhang, J.: A blockchain-based supply chain quality management framework. ICEBE **2017**, 172–176 (2017)
10. Szabo, N.: Formalizing and securing relationships on public networks. First Monday **2**(9), (1997)

11. Wohrer, M., Zdun, U.: Smart contracts: security patterns in the ethereum ecosystem and solidity. In: IWBOSE@SANER 2018, pp. 2–8. IEEE (2018)
12. Xu, Xiwei, Weber, Ingo, Staples, Mark: Case study: originChain. In: Architecture for Blockchain Applications, pp. 279–293. Springer, Cham (2019). https://doi.org/10.1007/978-3-030-03035-3_14
13. Zhu, Z., Qi, G., Zheng, M., Sun, J., Chai, Y.: Blockchain based consensus checking in decentralized cloud storage. Simul. Model. Pract. Theor. **102**, 101987 (2020). Special Issue on IoT, Cloud, Big Data and AI in Interdisciplinary Domains

Blockchain as a Countermeasure Solution for Security Threats of Healthcare Applications

Mubashar Iqbal$^{(\boxtimes)}$ (ID) and Raimundas Matulevičius$^{(\boxtimes)}$ (ID)

Institute of Computer Science, University of Tartu, Tartu, Estonia
{mubashar.iqbal,raimundas.matulevicius}@ut.ee

Abstract. Healthcare industry is digitising its healthcare operations and generating huge amounts of sensitive medical data to make prompt and informed decisions in patients' health diagnosis and care. The healthcare industry is subjected to a variety of security threats *(e.g., data tampering, theft, and counterfeit drugs)*. Blockchain is gaining traction to address such security threats and improve data integrity by turning healthcare operations into decentralised, transparent, and immutable manners. However, there is conceptual ambiguity and semantic gaps about blockchain as a countermeasure solution for healthcare security threats. In this work, we use the web ontology language to create blockchain-based security ontology (HealthOnt) to remove conceptual ambiguity and semantic gaps. The HealthOnt offers coherent and formal information models that present blockchain as a countermeasure solution for security threats of traditional healthcare applications.

Keywords: Blockchain · Healthcare applications · Security threats · Blockchain countermeasures · Security risk management

1 Introduction

Healthcare applications are integrating technology infrastructure to empower patients and the entire healthcare sector. The change facilitates the healthcare sector to make more prompt and informed decisions using digital medical data. The medical data is sensitive, confidential, and indispensable that plays an essential role in patients' health diagnosis, treatment and reduces medical mistakes. The growing medical data heighten the concerns to make it secure against various security threats (e.g., data tampering, theft, counterfeit drugs). Blockchain technology is emerging in healthcare to overcome such security challenges, enhance data integrity, and transform the transacting process into a decentralised, transparent, and immutable manner. For example, the study [1] presents the blockchain-based healthcare application along with cloud computing to protect medical data from being tampered, theft, and unauthorised use.

Blockchain is a decentralised, distributed, and immutable ledger technology that operates over a peer-to-peer (P2P) network [2]. A ledger contains a certain

© Springer Nature Switzerland AG 2021
J. González Enríquez et al. (Eds.): BPM 2021, LNBIP 428, pp. 67–84, 2021.
https://doi.org/10.1007/978-3-030-85867-4_6

and verifiable record of every single transaction ever made [1]. Blockchain technology is making inroads to various sectors, with the healthcare sector leading the way [3]. The success of blockchain-based applications is contingent on the medical data being accurate, verifiable, and untampered.

In healthcare, data is one of the most valuable assets. Healthcare applications suffer from various security threats [4–6] that could negate the confidentiality, integrity, and availability of medical data. The tampered medical data can cause major issues during the patient treatment process. Also, digital health records increase the risk of unauthorised access, information disclosure, and various internal and external threats. The study [7] investigated the security of healthcare applications, and findings reveal that organisations do not adhere to best practices when designing healthcare applications. Moreover, the technology infrastructure is incompatible with providing security measures by design.

The advent of blockchain technology has opened several research areas within the healthcare sector to preserve medical data, ensure data integrity, patient ownership to his data, easy exchange of medical data, and seamless medical insurance claims. However, there is conceptual ambiguity and semantic gaps about blockchain as a countermeasure solution for traditional healthcare applications [1,8,9]. Therefore, we build an ontology by investigating the security threats of traditional healthcare applications and how these security threats could be mitigated by utilising blockchain. The contribution of this work is twofold:

– A framework that explains the security threats and blockchain-based countermeasures to secure healthcare applications
– The construction of blockchain-based healthcare security ontology *(HealthOnt)*

We follow the security risk management (SRM) domain model [10,11] and develop a framework to explore the security threats of traditional healthcare applications. The framework assists us in building a blockchain-based healthcare security ontology (HealthOnt). The HealthOnt could support the selection of blockchain to security experts when designing healthcare applications. Also, the HealthOnt encodes traditional healthcare applications' information security into a dynamic ontology-based knowledge that can be extended, reused, or integrated with other security ontologies. The paper is structured as follows: Sect. 2 discusses the blockchain, research method, and related work. Section 3 presents the security risk analysis of healthcare applications and blockchain as a countermeasure solution. Section 4 gives an overview of ontology development. Section 5 is ontology evaluation, and Sect. 6 concludes the paper.

2   Background

2.1   Blockchain

Blockchain creates a chain of blocks and removes trusted intermediaries from the transaction process. A unique cryptographic hash links each block to the one before it. Blockchain could be classified as permissionless (e.g., Ethereum) or permissioned (e.g., Hyperledger Fabric). A permissionless blockchain is fully decentralised and accessible to everyone. Contrarily, a permissioned blockchain is partially decentralised with restrictions on who can join and access the operations. Blockchain comprises consensus mechanisms (e.g., Proof of Work (PoW), Proof of Stake (PoS)) to maintain the ledger state. Smart contract in blockchain is a piece of code that autonomously executes when certain conditions meet. Smart contract eliminates trusted intermediaries, less human intervention, reduces enforcement cost, prevents malicious or unintentional security threats [12].

Table 1. Blockchain features

Feature	Detail
Immutability	Once a record is added to the blockchain, it cannot be altered or deleted
Decentralised	Blockchain does not have a centralised single governing authority or a person. A group of distributed nodes maintains the network
Distributed	Operates over P2P network, and the participants' nodes have the same power in the network that share distributed computational power
Consensus	Helps to maintain the state and immutability of the ledger
Provenance	Each activity is recorded on a blockchain that lets everyone verify its authenticity
Tamper-evident	Blockchain detects any interference/tampering with the content
Cryptography	Allows blocks to be securely connected, ensuring consistency and immutability of the data stored in the blockchain
Distributed Access control	Blockchain-based access control provides decentralised and distributed resource authorisation
Permissioning	Categorisation of certain actions to be performed only by certain participants
Pseudonymous	Blockchain masks the user identity to not contain any identifiable information

Blockchain has a variety of features (Table 1) that makes it an irresistible and emerging technology in various applications domains. The features bring transparency, trust and tamper-resistance characteristics that are pillars of making the business and transactional procedures more secure, efficient and effective.

2.2 Research Method

This paper aims to present an ontological framework based on the SRM domain model to show blockchain as a countermeasure to mitigate various security threats of traditional healthcare applications. In this case, a systematic literature review (SLR) is appropriate since it allows the systematic analysis of relevant literature. We followed the review guidelines of Kitchenham [13] and specified the review protocol[1] to identify relevant papers and conduct this study.

The SRM domain model [10,11] helps us to structure the knowledge of blockchain as a countermeasure solution. Among other SRM approaches [15], the

---

[1] SLR settings are available in [14].

SRM domain model fulfils the criteria of ISO/IEC 27001 standard and explore three aspects *(e.g., assets-, risk-, and risk treatment-related)* during the early phases of information system development. The asset can be a system or business asset. The business asset has value and the system asset supports it. Security criteria (confidentiality - C, integrity - I, and availability - A) distinguish the security needs. The risk combines a risk event and impact. The risk event constitutes the threat and one or more vulnerabilities. The threat targets the system asset and exploits the vulnerability. The vulnerability is connected to the system assets and depicts their weaknesses. Impact harms the business asset and negates the security criteria. The risk treatment implements the security requirements as countermeasures to improve the system security.

2.3   Related Work

The research direction to secure healthcare applications by using blockchain is emerging. A few studies evaluated different security aspects of traditional healthcare applications and the role of blockchain to mitigate them.

Saha et al. [1] review the blockchain-based healthcare solutions to protect from data tampering and data leakage. The study presents a comparative analysis of different literature studies of healthcare applications that use blockchain. The survey [8] addresses the security and privacy concerns in healthcare. The authors explore the timeline of security attacks on medical data and various traditional security algorithms to defend against them. The traditional security algorithms are shown to be ineffective, and blockchain is used as an advanced architecture for the safe and secure execution of medical transactions and to maintain the security and privacy of digital medical records.

The study [9] describes the fundamental principles of blockchain to address the security and privacy issues of traditional healthcare applications. The study also discusses the technical advantages of blockchain in healthcare (e.g., faster and easier interoperability). The [16] presents the different use cases to address security and interoperability challenges of traditional healthcare applications.

Chukwu et al. [17] perform a SLR to explore the trust, security and privacy constraints of traditional digital health records and how blockchain plays a role to overcome them. The study evaluated 61 articles to address the traditional healthcare applications security challenges and blockchain-enabled emerging trends in healthcare research. The SLR [18] investigates the use cases and security challenges, including how blockchain can protect medical data from potential data loss, corruption or intentional security attacks. Jin et al. [12] present blockchain in healthcare for secure and privacy-preserving medical data sharing. The study argues that blockchain's tamper-evidence and decentralisation features could help build a secure medical data-sharing network.

The related works discuss the specific security aspect without addressing vulnerabilities, assets to protect, blockchain features, and not following any SRM domain model. In contrast, our study accumulates the security threats of traditional healthcare applications and how blockchain acts as a countermeasure solution to mitigate them. Moreover, we utilise the SRM domain model and develop an ontological framework that provides a dynamic knowledge base to facilitate the security of healthcare applications using blockchain.

3   Security Risk Analysis of Healthcare Applications

We analyse the literature studies using the SRM domain model to build a framework (Table 2) that presents the security threats, their vulnerabilities, assets to protect, and blockchain-based countermeasures. In this section, we only discuss the first five security threats from Table 2. The remaining security threats *(e.g., single-point failure, repudiation, insurance frauds, clinical trial fraud, tampering device settings, social engineering)* are discussed in [14].

3.1   Data Tampering

The traditional approaches lack control over data security, which is a major concern for healthcare organisations because it can put patient's lives at risk.

Vulnerabilities: In traditional healthcare applications, the access control is managed by a designated authority/individual that could be error-prone. The *weak centralised access control* [4,19–21] describes a case when the healthcare application fails to restrict unauthorised access to the resources. The attacker compromises the security and performs unauthorised actions that negate the integrity of medical records and confidentiality of patient data. The attacker uses unauthorised access to gain elevated privileges, execute commands, or bypass the security mechanisms to tamper with medical data. Moreover, traditional healthcare applications often rely on manual techniques or third-party providers to perform data verification and validation. These techniques *lack the proper mechanisms to verify and validate the authenticity of data* [2,22,23]. Consequently, the attacker can submit malicious content that the system can process and negate the integrity of medical records and confidentiality of patient data.

Countermeasures: Blockchain allows smart contracts-based distributed access control [24] that could regulate the users access to stored medical data. The system authenticates and identifies associated users according to their access rights deployed in a decentralised and distributed environment. Also, strong cryptographic primitives (e.g., attribute-based encryption) [25] help to build fine-grained access control. The records are difficult to modify/delete because of the ledger redundancy and append-only structure [26]. The healthcare applications on permissionless blockchain use PoW consensus to verify the executed transaction and data validation without requiring a third party before saving on the ledger [22]. Moreover, using the SHA-256 hashing algorithm, blockchain computes a unique hash id of original data that can be used to verify the authenticity of data [21]. Hyperledger fabric uses trusted authorised nodes to verify and validate the authenticity of data [2]. Blockchain is tamper-evident [20,21] and thus detects any unauthorised modifications. Blockchain builds strong audit trails in immutable ledgers by keeping a log of each performed action [20] over time that could be used to verify and validate the authenticity of data.

3.2   Data Theft

In healthcare, data theft has been on the rise over the past ten years, in 2020 reported 642 data thefts incidents [5] only in the United States.

Vulnerabilities: Databases are one of the most compromised assets [8] and centralised databases have *improper security controls* to protect against insider or outsider threats [25, 26]. The threats imposed by this vulnerability include: i) abuse of elevated privileges, ii) unauthorised access, iii) backup storage exposure, iv) database injection, v) default database accounts and configurations,

Table 2. Security risk analysis of traditional healthcare applications

Risk-related concept		Asset-related concept		Risk treatment concept	
Threat	Vulnerability	System asset	Business asset	Countermeasure	BC feature
Data tampering	Weak centralised access control mechanism	Healthcare database, Access control	Medical records (I), Patient data (C)	Distributed access control mechanism	
				Access control with cryptographic primitives (e.g., attribute-based encryption)	Access control
	No mechanism to verify and validate the authenticity of data	Healthcare database, Medical transactions	Medical records (I), Patient data (C), Data validation (I, A)	Distributed (shared) and append-only ledger	Distributed
				Proof of work-based consensus mechanism	Consensus
				Data validation without requiring third party	Consensus
				Unique hash id of original data	Cryptography
				HLF-based trusted authorised nodes	Permissioning
				Decentralised and tamper-resistant	Decentralised & Tamper-evident
				Immutable logging and data provenance	Provenance
Data theft	Improper security controls for centralised database	Healthcare system, Data access right	Healthcare database (I), Medical records (C)	Blockchain-based P2P network	Distributed
				Voting process to determine data access	Consensus
				Permissioned settings to restrict data access	Permissioning
				Access control with cryptographic primitives	
	Weak centralised access control mechanism	Access control	Medical records (C)	Distributed access control mechanism to control data leak	Access control
	No proper cryptographic controls	Healthcare system	Medical records (C)	Encrypts data and store on/off chain	Cryptography
				Store the encrypted and obfuscted data	
Medical records mishandling	Patients have weak control over their medical records	Data access right	Medical records (I, C)	Blockchain enables patients to control the access to their data	Permissioning
	Relying on a third-party			Data validation without requiring third party	Decentralised
	No guarantee of electronic medical records authenticity	Healthcare database	Medical records (C)	Decentralised and tamper-resistant	Decentralised & Tamper-evident
				Consensus mechanism	Consensus
Counterfeit drugs	Weak traceability controls in pharmaceutical supply chain	Drugs details, Supply chain	Drug traceability (I)	Immutable and traceable drug trails	Provenance & Immutability
Man in the middle attack	Weak controls to secure communication	Network, Data exchange	Communication (I)	Distributed IPFS for storage	Distributed & Cryptography
				P2P-based encrypted communcation	
	Lack of anonymisation of patient medical records	Healthcare system	Medical records (I, C)	Blockchain anonymise the data	Pseudo-anony mous
Single point failure	Relying on centralised server Weak implementation to handle large number of requests	Healthcare database and system	Server (A), Services (A)	Decentralised distributed P2P network	Decentralised & Distributed
Repudiation	Weak controls to prove illegal data changes by authorised users	Healthcare system	Medical records (I)	Blockchain-based versioning scheme to track each performed operation	Provenance & Immutability
	Lack of immutable logs	Action logs	Medical records (I)	Immutable log of all performed activities	
Insurance fraud	No proper authenticity to verify the insurance claim	Medical bills, Insurance data	Insurance claim (I)	Decentralised verification of insurers	Permissioning
				Verified records are distributed among nodes	Distributed
Clinical trial fraud	Inadequate clinical trials data	Clinical trial data, Data access right	Data processing (I, C)	Distributed nature and use of cryptography	Cryptography
	Improper patient recruitment and lack of data access			Blockchain provides data ownership	Permissioning
				Data saved on blockchain cannot be altered	Immutability
Tampering device settings	Weak controls on settings of medical devices	IoT devices	Device settings (I, A)	Storing devices settings in distributed immutable ledger	Immutability
Social engineering	Possible to manipulate employess to get data access	Employees, Stakeholders	Medical records (I)	Only relevant employees have access to particular information or part of information	Permissioning

vi) malware and the vii) human factor [5,8]. Overall it negates the integrity of the healthcare system and confidentiality of medical data.

Similar to data tampering, the attacker can steal medical data due to *weak centralised access control* [25,27] that leads the attacker to gain unauthorised access, elevated privileges, or bypass security mechanisms. As a result, it negates the confidentiality of medical data. Traditional healthcare applications use cryptography to save data securely and achieve information security objectives. However, it *lacks cryptographic control* [27] over data since the centralised authority/individual is responsible for the administration of the database (keeping elevated privileges, encryption/decryption keys). If the security of the system is compromised, then the attacker can steal the medical data.

Countermeasures: Blockchain works on a P2P-based distributed network where nodes behave both as a server and client to send and receive data directly with each other. This mechanism helps to protect the data leakage to unauthorised network users [2]. The solution proposed in [26] uses the voting process (e.g., QuorumChain algorithm) to determine which nodes are allowed to access certain types of data. The permissioned blockchains define permission settings to restrict data access only to authorised nodes [21,27]. Similar to data tampering countermeasures, the strong cryptographic primitives (e.g., attribute-based encryption) [25] and smart contracts-based distributed access control mechanism [22] allows only authorised users to access medical data. The Ancile framework [26] uses the proxy re-encryption to encrypt the data and store hashes data on/off-chain, [25] suggests data obfuscation to protect data on/off-chain.

3.3 Medical Records Mishandling

Healthcare staff must ensure that medical records are kept private and safe. But medical records mishandling is one of the common HIPAA violations [6].

Vulnerabilities: The medical institutions control and manage the patient's medical data where the non-relevant individuals can access it as well. Thus, patients have *weak control over their medical records* [28]. Also, the patient is unaware of how his data is processed or with whom it is shared. In some cases, the individual from a medical institution involves in illegal medical data trade [29] that negates the integrity and confidentiality of medical records.

Hospitals and healthcare applications *rely on third-party* [23,25] vendors (e.g., IT vendors, pharmacies, insurance companies, etc.) daily to perform their routine functions. These third-party vendors have access to the patient's medical data. They could intentionally sell medical data to data brokers or become a source of data breach and negate the confidentiality of medical data.

Also, the medical data is managed by a designated authority/individual, and the system administration governs the system with elevated privileges. If any such point is compromised, the attacker can manipulate and negate the integrity of medical data without leaving the traces. Therefore, traditional healthcare applications *cannot guarantee the authenticity* [23] of digital medical records.

Countermeasures: The permission settings and distributed access control enable patients to handle their medical data [19,30]. The blockchain performs data validation during the consensus process before saving on the ledger. For example, blockchain provides a transparent platform to define data validation rules which are agreed upon by decentralised and distributed network nodes [31]. Then, all the nodes follow those rules to validate the data. The blockchain-based applications detect and discard all the unauthorised changes [30] if the majority of the network is honest (e.g., the adversary does not control 51% computing power). This process establishes a tamper-resistant environment [32].

3.4   Counterfeit Drugs (Fake Medicine)

For years, the pharmaceutical supply chain has been struggling to monitor its products and avoid fake medicine. According [19,28], 10–30% (worth $200 billion) of drugs sold worldwide each year are counterfeit.

Vulnerabilities: Counterfeit drugs are on the rise, posing significant health risks. In pharmaceuticals, after manufacturing, drugs are moved from production stocks to wholesale distributors, which then move to retail firms. Customers purchase drugs from retailers. Due to *weak traceability controls* (e.g., ineffective data sharing, no traceable records) [3,19,28] in the pharmaceutical supply chain, there is a risk of fake medicines being introduced during this process.

Countermeasures: Blockchain offers a solution to enable pharmaceutical traceability, real-time access to data and supply chain validation by creating a log to track each step [3,19,28]. For example, IBM Research uses blockchain to reduce or eliminate the drug counterfeiting problems in Kenya [28] by using immutable and traceable logs at each stage of the pharmaceutical supply chain.

3.5   Man in the Middle (MitM) Attack

MitM attacks are rising in healthcare systems to gain sensitive information [33].

Vulnerabilities: The attacker can exploit the *weak controls of secure communication* [4] in traditional healthcare applications and negate the integrity of communication assets. For example, not properly implementing (or having) cryptographic functionality or lack of fine-grained access control mechanism. Moreover, due to *lack of anonymisation of patient medical records* [34] the medical data is associated directly with patient identity. The attacker can get the data to trigger a ransomware attack, publish it online or deny access to it.

Countermeasures: The authors [4] introduce the distributed interplanetary file system (IPFS) for storage along with blockchain and blockchain-based data encryption to reduce communication and computation overhead that establish a secure communication channel. Blockchain works on a P2P-based distributed network where nodes behave both as a server and client to exchange encrypted data directly with each other. This feature of blockchain makes it hard for an attacker to intercept communication or data analysis/sniffing [2, 19]. Blockchain maintains pseudo-anonymity, the patients and their medical data is linked with a public address. Also, the data processing on a blockchain is anonymous [19] and blockchain anonymises the medical data to hide the actual identity [34].

4   Healthcare Security Ontology

Ontology elaborates the meaning of concepts within a domain to overcome the consequences of a misunderstanding. The study [35] illustrates the reasons that motivate the development of an ontology. For instance, ontology makes it possible to i) share a common understanding, ii) reuse of domain knowledge, iii) make domain assumptions explicit, iv) separate domain and operational knowledge, v) analyse domain knowledge.

HealthOnt is based on web ontology language (OWL) and WWW Consortium (W3C). OWL is a semantic web language to illustrate rich and complex knowledge about things and their relations [36]. We use SPARQL (SPARQL Protocol and RDF Query Language) as a semantic query language [37] to get results from an ontology. We utilise the **ontology construction method** [38] that has five stages: i) Identify purpose & scope, ii) Building ontology, it includes capture, coding and integrating phases, iii) evaluation, iv) documentation, and v) guidelines. In [39], we follow the same ontology construction method to explore and build an ontology for security threats of Corda-based financial applications.

Scope and Purpose: The instructions provided in [35] help us to define the scope and purpose of our ontology. The purpose is to build a knowledge base of blockchain-enabled countermeasures for healthcare applications. The scope covers the *domain of ontology* (e.g., blockchain as a countermeasure solution), *use of ontology* (e.g., SRM of healthcare applications), *questions that ontology answers* (e.g., what assets to protect, what are the threats, vulnerabilities, and countermeasures), *who will maintain the ontology?* (e.g., security experts).

Building Ontology: We use Protege to capture and code domain knowledge (e.g., concepts and their relationships) into taxonomic classifications. The classifications refine the concepts belonging to assets, security threats, vulnerabilities, countermeasures, and blockchain features associated with countermeasures.

Assets Classification: Assets are classified as business and system assets (Fig. 1). Security criteria is a constraint of business assets, and system assets support business assets. For example, business asset "MedicalRecord" has Constraint Integrity, and System assets "AccessControl" supports "MedicalRecord".

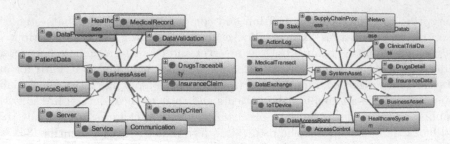

Fig. 1. Business and system assets classification

The asset class definition explains sub-classes (e.g., BusinessAsset and SystemAsset) and restrictions "hasConstraint" and "supports".

```
Class (Asset SubClass (BusinessAsset SystemAsset)
    BusinessAsset hasConstraint someValuesFrom (SecurityCriteria)
    SystemAsset supports someValuesFrom (BusinessAsset)
)
```

Threats Classification: Security threats classification (Fig. 2) is built upon the threats that are mitigated using blockchain. In traditional healthcare applications, security threats exploit vulnerabilities and target some system asset(s). Threat class has a restriction "exploits" on someValuesFrom the Vulnerability. Another restriction "targets" on someValuesFrom the SystemAsset.

```
Class (Threat SubClass ( DataTampering DataTheft .... )
    restriction ( exploits someValuesFrom (Vulnerability) )
    restriction ( targets someValuesFrom (SystemAsset) )
)
```

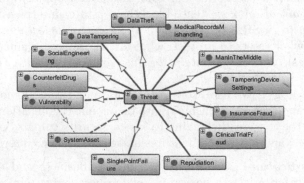

Fig. 2. Security threats classification

For example, in traditional healthcare applications, the attacker can trigger "DataTampering" threat by exploiting a vulnerability "ErrorProneAuthenticityOfData" or "WeakAccessControl". The "DataTampering" threat targets the system assets (e.g., AccessControl, HealthcareDatabase, or MedicalTransaction).

Vulnerabilities classification: This classification (Fig. 3) is built upon the weaknesses in healthcare applications that enable some security threats.

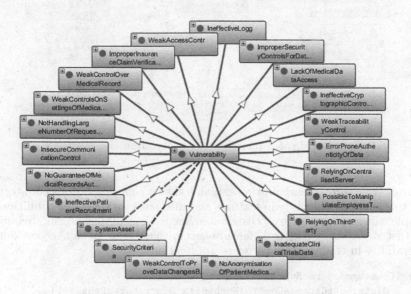

Fig. 3. Vulnerabilities classification

The vulnerability class definition explains various vulnerabilities that are characteristic of system assets and negates the security criteria of business assets.

```
Class (Vulnerability SubClass ( WeakAccessControl .... )
    restriction ( negates someValuesFrom (SecurityCriteria) )
    restriction ( characteristicOf someValuesFrom (SystemAsset) )
)
```

For example, the weak implementation of access control presents a weakness that the attacker can exploit and get unauthorised access. A vulnerability "WeakAccessControl" is a characteristicOf "SystemAsset" (AccessControl, HealthcareDatabase) and negates some (Integrity or Confidentiality).

Countermeasures Classification: Countermeasures classification (Fig. 4) presents the counteract that mitigates the vulnerabilities and improves the security of the system. The countermeasures belong to the blockchain features.

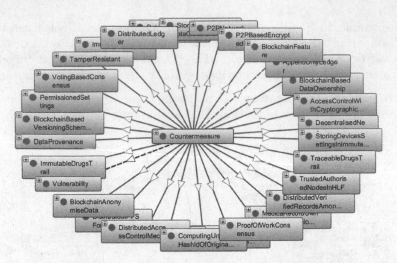

Fig. 4. Countermeasures classification

Countermeasure class definition explains that it contains various countermeasures that belong to blockchain features and mitigates various vulnerabilities. For example, the countermeasure "DistributedAccessControlMechanism" belongs to the "DistributedAccessControl" feature and mitigates the "WeakAccessControl" vulnerability in traditional healthcare applications.

```
Class (Countermeasure SubClass (
     DistributedAccessControlMechanism DistributedLedger ....
  )
  restriction ( belongsTo someValuesFrom (BlockchainFeature) )
  restriction ( mitigates someValuesFrom (Vulnerability) )
)
```

Blockchain Feature Classification: Blockchain features classification (Fig. 5) presents the characteristics that are associated with blockchain-based countermeasures (Table 1). For example, the countermeasure "DistributedAccessControlMechanism" belongs to the "DistributedAccessControl" blockchain feature.

```
Class (BlockchainFeature SubClass (
     DistributedAccessControl Immutability Provenance ....
  )
)
```

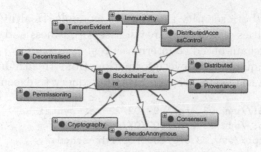

Fig. 5. Blockchain features classification

Documentation and Guidelines: HealthOnt is accessible online (Table 3) and we use the Protege annotations to document the concepts & relationships. Also, we utilise the Protege OntoGraf plugin to generate classifications graphs and the Pellet reasoner to validate the consistency of our ontology.

Table 3. HealthOnt resources

Resource	Resource URL
HealthOnt	https://mmisw.org/ont/~mubashar/HealthOnt
GitHub	https://github.com/mubashar-iqbal/HealthOnt
Protege	https://protege.stanford.edu/
OntoGraf	https://protegewiki.stanford.edu/wiki/OntoGraf
Pellet Reasoner	https://protegewiki.stanford.edu/wiki/Using_Reasoners

5 Ontology Evaluation

We use the task-based [40] evaluation technique. This technique allows learning about HealthOnt applicability. The efficient evaluation technique contributes to the scientific value of an ontology. For instance, consider healthcare security experts working on a healthcare application or a healthcare organisation looking for viable solutions to address the security threats associated with medical data tampering and theft. Due to the conceptual ambiguity and semantic gaps, both the healthcare security experts and organisation are unaware of the blockchain's countermeasures that could mitigate both security threats.

In this case, HealthOnt supports the selection of blockchain to mitigate both security threats of traditional healthcare applications and determine what assets to protect. Also, HealthOnt can assist the conceptual design and technological implementation of both security threats. For example, HealthOnt helps to identify the vulnerabilities of security threats, assets (business and system assets) to protect, and blockchain-based countermeasures.

The SPARQL queries retrieve information from HealthOnt. The following code is required in the header of each SPARQL query to execute it successfully.

```
PREFIX rdf: <http://www.w3.org/1999/02/22-rdf-syntax-ns#>
PREFIX owl: <http://www.w3.org/2002/07/owl#>
PREFIX rdfs: <http://www.w3.org/2000/01/rdf-schema#>
PREFIX xsd: <http://www.w3.org/2001/XMLSchema#>
PREFIX HealthOnt: <https://mmisw.org/ont/~mubashar/HealthOnt#>
```

The SPARQL **Query #1** retrieves the security threats related to data tampering and theft, their vulnerabilities, and assets to protect. The query compiles results based on the defined relationships *(exploits and targets)*. For example, **Threat** (DataTampering) *exploits* **Vulnerability** (WeakAccessControl), and **Vulnerability** (WeakAccessControl) *targets* **System Asset** (AccessControl or HealthcareDatabase or MedicalTransaction). Similar results for DataTheft and other security threats (reference Table 2).

```
Query# 1 SELECT ?Threat ?Vulnerability ?SystemAsset WHERE {
    ?Threat rdfs:subClassOf ?Vulnerability .
    ?Threat rdfs:subClassOf ?SystemAsset .
    ?Vulnerability owl:onProperty HealthOnt:exploits .
    ?SystemAsset owl:onProperty HealthOnt:targets .
    ?Threat rdfs:label ?FilterByThreat .
    FILTER regex(?FilterByThreat, "DataTampering|DataTheft") .
}
```

The SPARQL **Query #2** retrieves the countermeasures and vulnerabilities that are associated with data tampering and theft. The query fetch results based on the relationship *(mitigates)*. For example, **Countermeasure** (DistributedAccessControl) *mitigates* **Vulnerability** (WeakAccessControl). Similar results for DataTheft and other security threats (reference Table 2).

```
Query# 2 SELECT ?Countermeasure ?Vulnerability WHERE {
    ?Countermeasure rdfs:subClassOf HealthOnt:Countermeasure .
    ?Countermeasure rdfs:subClassOf ?Vulnerability .
    ?Vulnerability owl:onProperty HealthOnt:mitigates .
    ?Countermeasure HealthOnt:Mitigates ?FilterByThreat .
    FILTER regex(?FilterByThreat, "DataTampering|DataTheft") .
}
```

Ontology validation is an important part to ensure the correctness of ontological knowledge and ontological reasoning meaning. We perform the qualitative assessment to validate the HealthOnt using the criteria of [40, 41] (Table 4).

Table 4. HealthOnt validation

Criteria	Detail
Accuracy	We utilise the scientific literature to define classes, properties, and individuals
Adaptability	HealthOnt provides a conceptual foundation for a range of anticipated tasks (e.g., threats, vulnerabilities, and countermeasures)
Clarity	Definitions related to HealthOnt concepts and relationships are documented
Completeness	SRM domain model enables the richness and granularity of HealthOnt
Computational efficiency	We use the Pellet reasoner to process the HealthOnt, and SPARQL for querying results. Pellet reasoner and SPARQL are fast and computationally efficient tools
Conciseness	HealthOnt includes only essential terms and explains weak points in threats to validity
Consistency	We use the Pellet reasoner to check HealthOnt consistency and avoid contradictions in ontology concepts and relationships
Organisational fitness	We follow the well-defined ontology construction method. HealthOnt is available online, and it can be extended, reused, or integrated with other security ontologies

6   Discussion and Concluding Remarks

In this work, we utilise the SRM domain model to build HealthOnt by exploring traditional healthcare applications security threats. We define the scope of our ontology and develop the classifications related to assets, security threats, vulnerabilities, countermeasures, and blockchain features. HealthOnt is publicly available and encodes the information into a dynamic ontology-based knowledge that can be extended, reused or integrated with other security ontologies. HealthOnt can support the iterative process of SRM and it is possible to update HealthOnt continuously when new security threats or countermeasures emerge. To assure the quality of empirical studies, we extended our discussion by overviewing the future work and threats to validity [42].

Future Work: During our research, we discover that the blockchain is considered to be a security tool for healthcare applications [21,28]. However, blockchain-based healthcare applications are not completely proven to protect data from various security threats. In fact, there are a number of ways to negate the security of blockchain-based applications [43,44]. In future work, we are extending HealthOnt by including the *security threats that could appear* in blockchain-based healthcare applications. Also, *ontology validation requires further work* to determine whether the correct ontology was developed and whether the ontology accurately models the real world for which it was developed. In correspondence, we will use the real-world case of traditional healthcare application to validate and demonstrate the applicability of HealthOnt. Furthermore, an evaluation based on ontology domain experts is required to perceive the significance of HealthOnt contribution, to derive what is missing, and to determine HealthOnt accuracy, comprehensiveness, and technical correctness.

Threats to Validity: We address the threats to validity by [42] mapping. The relevant threats are restricted time span, publication bias, subjective interpretation, and lack of expert evaluation. The *restricted time-span* is that the researcher cannot predict other applicable studies beyond the time span. For example, blockchain is relatively new but continuously evolving. Therefore a wide variety of countermeasures will emerge in the future. The **publication bias** is that the related studies are more likely to report positive results than negative results. The threat of **subjective interpretation** exists since we might have different interpretations and opinions related to identified threats, vulnerabilities, and countermeasures. Moreover, a **lack of expert evaluation** may also lead to a subjective interpretation and erroneous conclusion. These threats and no validation using a real-world case of healthcare application raise the concerns related to **weak evaluation of ontology**. For instance, errors or limitations may be discovered when applying HealthOnt in real-world industrial settings. However, in our future work, we are focusing on overcoming these threats.

Acknowledgement. This work was partially supported by the ERASMUS+ sectoral alliance program on *A Blueprint for sectoral cooperation on blockchain skill and development (CHAISE)*, under grant no. 621646-EPP-1-2020-1-FR-EPPKA2-SSA-B. The European Commission support for the production of this publication does not constitute endorsement of the contents which reflects the views only of the authors, and the Commission cannot be held responsible for any use which may be made of the information contained therein.

References

1. Saha, A., Amin, R., Kunal, S., Vollala, S., Dwivedi, S.K.: Review on Blockchain technology based medical healthcare system with privacy issues, pp. 1–14 (2019)
2. Chen, J., Ma, X., Du, M., Wang, Z.: A blockchain application for medical information sharing. In: TEMS-ISIE, pp. 1–7 (2018)
3. Narikimilli, N.R.S., Kumar, A., Antu, A.D., Xie, B.: Blockchain applications in healthcare – A review and future perspective. In: Chen, Z., Cui, L., Palanisamy, B., Zhang, L.-J. (eds.) ICBC 2020. LNCS, vol. 12404, pp. 198–218. Springer, Cham (2020). https://doi.org/10.1007/978-3-030-59638-5_14
4. Xu, J., et al.: Healthchain: a blockchain-based privacy preserving scheme for large-scale health data. IEEE Internet Things J. **6**(5), 8770–8781 (2019)
5. HIPAA: 2020 Healthcare Data Breach Report: 25% Increase in Breaches in 2020 (2021). https://bit.ly/3uN16BN
6. Zabel, L.: 10 common HIPAA violations and preventative measures to keep your practice in compliance (2016). https://bit.ly/34E8Hrx
7. Mansfield-Devine, S.: Your life in your hands: the security issues with healthcare apps. Network Secur. **4**(2016), 14–18 (2016)
8. Hathaliya, J.J., Tanwar, S.: An exhaustive survey on security and privacy issues in Healthcare 4.0. Comput. Commun. **153**(2020), 311–335 (2020)

9. Linn, L.A., Koo, M.B.: Blockchain for health data and its potential use in health it and health care related research. In: HealthIT.gov, pp. 1–10 (2014)
10. Dubois, É., Mayer, N., Heymans, P., Matulevičius, R.: A systematic approach to define the domain of information system security risk management. In: Nurcan, S., Salinesi, C., Souveyet, C., Ralyté, J. (eds.) Intentional Perspectives on Information Systems Engineering, pp. 289–306. Springer, Heidelberg (2010). https://doi.org/10.1007/978-3-642-12544-7_16
11. Matulevičius, R.: Secure system development. In: Fundamentals of Secure System Modelling, pp. 199–207. Springer, Cham (2017). https://doi.org/10.1007/978-3-319-61717-6_12
12. Jin, H., Luo, Y., Li, P., Mathew, J.: A review of secure and privacy-preserving medical data sharing. IEEE Access 7, 61656–61669 (2019)
13. Kitchenham, B., Charters, S.: Guidelines for Performing Systematic Literature Reviews in Software Engineering. EBSE Technical Report, Version 2.3 (2007)
14. Iqbal, M., Matulevičius, R.: Blockchain as a Countermeasure Solution for Security Threats of Healthcare Applications (Technical report) (2021). https://github.com/mubashar-iqbal/HealthOnt/blob/main/Technical_Report.pdf
15. Ganji, D., Kalloniatis, C., Mouratidis, H., Gheytassi, S.M.: Approaches to develop and implement ISO/IEC 27001 standard - Information security management systems: a systematic literature review. Int. J. Adv. Softw. 12(3), 228–238 (2019)
16. Randall, D., Goel, P., Abujamra, R.: Health and Medical Informatics Blockchain Applications and Use Cases in Health Information Technology, pp. 8–11 (2017)
17. Chukwu, E., Garg, L.: A systematic review of blockchain in healthcare: frameworks, prototypes, and implementations. IEEE Access 8, 21196–21214 (2020)
18. Agbo, C., Mahmoud, Q., Eklund, J.: Blockchain Technology in Healthcare: A Systematic Review. Healthcare, pp. 1–56 (2019)
19. Yaqoob, I., Salah, K., Jayaraman, R., Al-Hammadi, Y.: Blockchain for healthcare data management: opportunities, challenges, and future recommendations. Neural Comput. Appl. 1–16 (2021). https://doi.org/10.1007/s00521-020-05519-w
20. Bhuiyan, Z.A., Wang, T., Wang, G.: Blockchain and Big Data to Transform the Healthcare, pp. 2–8 (2018)
21. Han, H., Huang, M., Zhang, Yu., Bhatti, U.A.: An architecture of secure health information storage system based on blockchain technology. In: Sun, X., Pan, Z., Bertino, E. (eds.) ICCCS 2018. LNCS, vol. 11064, pp. 578–588. Springer, Cham (2018). https://doi.org/10.1007/978-3-030-00009-7_52
22. Hussein, A.F., ArunKumar, N., Ramirez-Gonzalez, G., Abdulhay, E., Tavares, J.M.R., Albuquerque, V.: A medical records managing and securing blockchain based system supported by a genetic algorithm and discrete wavelet transform. Cognitive Syst. Res. 52, 1–11 (2018)
23. Du, M., Chen, Q., Chen, J., Ma, X.: An optimized consortium blockchain for medical information sharing. IEEE Trans. Eng. Manag. 13, 1–13 (2020)
24. Francesco, D., Ricci, L., Mori, P.: Distributed access control through blockchain technology blockchain. In: ERCIM News: Blockchain Engineering, pp. 31–32 (2017)
25. Esposito, C., De Santis, A., Tortora, G., Chang, H., Choo, K.K.R.: Blockchain: a panacea for healthcare cloud-based data security and privacy? IEEE Cloud Comput. 5(1), 31–37 (2018)
26. Dagher, G.G., Mohler, J., Milojkovic, M., Marella, P.B.: Ancile: privacy-preserving framework for access control and interoperability of electronic health records using blockchain technology. Sustain. Cities Soc. 39, 283–297 (2018)

27. Al Omar, A., Rahman, M.S., Basu, A., Kiyomoto, S.: MediBchain: a blockchain based privacy preserving platform for healthcare data. In: Wang, G., Atiquzzaman, M., Yan, Z., Choo, K.-K.R. (eds.) SpaCCS 2017. LNCS, vol. 10658, pp. 534–543. Springer, Cham (2017). https://doi.org/10.1007/978-3-319-72395-2_49
28. Martino, F.D.D., Klein, S.D., Neil, J.O., Huang, Y., Nisson, L., Race, M.: Transforming the U. S. Healthcare Industry with Blockchain Technology, pp. 1–7 (2019)
29. Thielman, S.: Your private medical data is for sale - and it's driving a business worth billions (2017). https://bit.ly/3ceaacp
30. Shi, S., He, D., Li, L., Kumar, N., Khurram, M.: Applications of blockchain in ensuring the security and privacy of electronic health record systems: a survey. Comput. Secur. 101996 (2020)
31. Dexter, S.: How Are Blockchain Transactions Validated? Consensus VS Validation (2018). https://www.mangoresearch.co/blockchain-consensus-vs-validation
32. Tosh, D.K., Shetty, S., Liang, X., Kamhoua, C.A., Kwiat, K.A., Njilla, L.: Security implications of blockchain cloud with analysis of block withholding attack. In: 17th IEEE/ACM International Symposium CCGRID, pp. 458–467 (2017)
33. SpecOpsSoft: The countries experiencing the most 'significant' cyber-attacks (2020) https://bit.ly/3idba4m
34. Ali, M.S., Vecchio, M., Putra, G.D., Kanhere, S.S., Antonelli, F.: A decentralized peer-to-peer remote health monitoring system. Sensors (Switzerland) **20**(6), 1–18 (2020)
35. Noy, N.F., McGuinness, D.L.: Ontology development 101: a guide to creating your first ontology. In: Stanford Knowledge Systems Laboratory, pp. 1–25 (2001)
36. Group, O.W.: Web Ontology Language (OWL). https://www.w3.org/OWL
37. Herzog, A., Shahmehri, N., Duma, C.: An ontology of information security. IJISP **1**(4), 1–23 (2007)
38. Uschold, M., Gruninger, M.: Ontologies: principles, methods and applications. Knowl. Eng. Rev. **11**(2), 93–136 (1996)
39. Iqbal, M., Matulevičius, R.: Corda security ontology: example of post-trade matching and confirmation. Baltic J. Modern Comput. **8**(4), 638–674 (2021)
40. Raad, J., Cruz, C.: A survey on ontology evaluation methods. In: HAL Archives-Ouvertes (2018)
41. Vrandečić, D.: Ontology evaluation. In: Staab, S., Studer, R. (eds.) Handbook on Ontologies. IHIS, pp. 293–313. Springer, Heidelberg (2009). https://doi.org/10.1007/978-3-540-92673-3_13
42. Zhou, X., Jin, Y., Zhang, H., Li, S., Huang, X.: A map of threats to validity of systematic literature reviews in software engineering, pp. 153–160 (2016)
43. Iqbal, M., Matulevičius, R.: Blockchain-based application security risks: a systematic literature review. BIOC **2019**, 1–26 (2019)
44. Iqbal, M., Matulevičius, R.: Exploring Sybil and double-spending risks in blockchain systems. IEEE Access **9**, 76153–76177 (2021)

Studying Bitcoin Privacy Attacks and Their Impact on Bitcoin-Based Identity Methods

Simin Ghesmati[1,2]([✉]), Walid Fdhila[1,3], and Edgar Weippl[1,3]

[1] SBA Research, Vienna, Austria
{sghesmati,wfdhila,eweippl}@sba-research.org
[2] Vienna University of technology, Vienna, Austria
[3] University of Vienna, Vienna, Austria

Abstract. The Bitcoin blockchain was the first publicly verifiable, and distributed ledger, where it is possible for everyone to download and check the full history of all data records from the genesis block. These properties lead to the emergence of new types of applications and the redesign of traditional systems that no longer respond to current business needs (e.g., transparency, protection against censorship, decentralization). One particular application is the use of blockchain technology to enable decentralized and self-sovereign identities including new mechanisms for creating, resolving, and revoking them. The public availability of data records has, in turn, paved the way for new kinds of attacks that combine sophisticated heuristics with auxiliary information to compromise users' privacy and deanonymize their identities. In this paper, we review and categorize Bitcoin privacy attacks, investigate their impact on one of the Bitcoin-based identity methods namely did:btcr, and analyze and discuss its privacy properties.

Keywords: Decentralized identifier · DID · Privacy · BTCR · Blockchain · Bitcoin

1 Introduction

Bitcoin blockchain [1] is an immutable tamper-proof distributed ledger, where addresses are used as pseudonyms (hashes of public keys), and eventually associated with amounts of bitcoins that can be redeemed using the corresponding private keys. Besides cyrptocurrencies, blockchain technology has enabled a large number of new applications that range from coordinating and monitoring cross-organizational business processes [2–4] to designing new methods for distributed identity management [5,6]. Business process automation, for example, requires that the different actors (e.g., customers, employees, business partners), resources, and services interact with each other in a trusted manner. This trustworthy communication, in turn, requires that entities can establish trusted communication channels, with certitude about the authenticity of the entities

J. González Enríquez et al. (Eds.): BPM 2021, LNBIP 428, pp. 85–101, 2021.
https://doi.org/10.1007/978-3-030-85867-4_7

they are interacting with. In this regard, identity continues to play a primordial role as an enabler of such trustworthy communications. Identity is a collection of data, which defines the attributes of a subject, e.g., cryptographic material for establishing communication (public key), verification methods for proving identity ownership, or service endpoints. Traditional systems often relied on isolated, centralized or federated architectures to manage identities. While in an isolated model each third party service/business is itself the identity provider IDP (i.e., responsible for storing and managing identities data), centralized and federated models both delegate identity management to separate IDPs, that work in isolation or federation, respectively. However, recent breaches (e.g., 500 million Facebook accounts, and 700 million LinkedIn accounts leaked[1]) exposed the limits of such systems and called for more decentralized models that give users control over their data. With the advent of blockchain, it became possible to create and resolve decentralized identifiers (DIDs) without having to rely on centralized authorities. This opened the door for a multitude of proposals (DID methods) that enable decentralized creation, resolution, update, and revocation of DIDs. It is noteworthy to point out that these DID methods rely on different blockchain technologies and architectural designs. One of the first proposed DID methods specifically use the Bitcoin blockchain and is called did:btcr [7].

In our previous research [8], we demonstrated how it is possible to combine sophisticated heuristics with auxiliary information (e.g. address tag databases) to correlate Bitcoin addresses with their corresponding real identities, which may put users' privacy at risk. In this paper, we review and categorize privacy attacks on the Bitcoin blockchain, which not only may reveal the links between addresses and real-world identities, but also correlate between different identities. Next, we address Bitcoin privacy attacks' impact on the DID method did:btcr. To this end, we adopted the terminology from RFC 6973 [9]. The contributions of the paper are in two folds: (i) Categorizing Bitcoin privacy attacks, and (ii) Investigation of the privacy issues in did:btcr.

The remainder of the paper is organized as follows: In Sect. 2, we describe the main concepts, while in Sect. 3 we introduce the methodology, categorize Bitcoin privacy attacks and explain how they may impact users' privacy. In Sect. 4, we investigate privacy issues in DiD BTCR method, and in Sect. 5, we conclude the paper and provide the future work.

2 Background

2.1 Bitcoin

In Bitcoin, transactions consist of input and output addresses. The input refers to the output of one of the previous transactions. A mining fee is often included as part of the transaction to increase its chance of being considered by miners. This explains why the sum of the inputs should always be larger than the sum of the outputs. Additionally, whenever the sum of the inputs plus the fee is

---

[1] https://haveibeenpwned.com/.

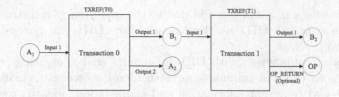

Fig. 1. Bitcoin transactions

larger than the amount that should be spent, a fresh address, namely a change address is created to send the remainder to the sender [10]. Figure 1 illustrates a simplified form of Bitcoin transactions. In the first transaction, Alice (A_1) sends the bitcoins to Bob (B_1) and gets the remainder back to her change address (A_2). In the following transaction, Bob sends the bitcoins from his address (B_1) to his another address (B_2), while additionally specifying an optional OP_RETURN output. OP_RETURN is an opcode that enables embedding a small amount of data within a transaction.

Bitcoin is a publicly available ledger, and therefore, all the transaction details including sender's and recipient's addresses, the values of transactions, and corresponding timestamps remain visible and can be checked by anyone. Despite the nice properties of blockchain, it in turn, created a niche for attackers to exploit such available data for malicious purposes. Previous and ongoing research has identified several privacy issues that can reveal identities and effectively find the relationships between Bitcoin addresses and the corresponding identities [11–14].

2.2   Decentralized Identifiers (DIDs)

Entities, including users and organizations, utilize global unique identifiers for a variety of use cases such as telephone numbers, ID numbers or URLs. These identifiers are often issued and managed by central authorities. Previous data breaches, however, diminished trust in such centralized architecture and called for decentralized management of identities, where users become their own identity providers. As a result, blockchain-based decentralized identifiers [15] have been proposed, which rely on blockchain and additional cryptographic techniques to prove identifiers' ownership without having to rely on a trusted entity.

A **decentralized identifier (DID)** is a string that includes three main parts: the scheme, the DID method, and the DID method identifier, which should be unique within the DID method. The syntax according to the W3C recommendation[2] is as follows.

Scheme : *DIDmethod* : *DID_method_identifier*

DIDs are usually associated with **DID documents**; i.e., documents that contain information about the verification methods (e.g. cryptographic public keys) and the service endpoints required to interact with the DID subjects. The DID subject is the entity that is identified by the DID, and can be a person,

---

[2]   https://www.w3.org/TR/2021/CRD-did-core-20210609/.

an object or an organization. In addition to the underlying infrastructure (e.g., Bitcoin, Ethereum), a **DID method** defines how DIDs are created, resolved, updated, and revoked.

While DIDs in conjunction with DID documents enable creating trustworthy communications (how to communicate with identity owners), **verifiable credentials (VCs)** represent information and claims about identity owners (e.g., name, age, diplomas) [16]. These credentials can be issued by different issuers (e.g., university, employer), and can be cryptographically verified by any third party without having to contact the corresponding issuers.

2.3   BTCR

The BTCR method [7] uses the Bitcoin blockchain to manage DIDs. In did:btcr, DIDs are created using the transaction references $TXRef$, only known once transactions are confirmed. The following is an example of a did:btcr (adopted from [7]), where did is the scheme part, btcr is the DID method part, and xyv2-xzpq-q9wa-p7t is the identifier which is the transaction reference. Transaction reference follows BIP 0136, which encodes transactions positions (including the chain, block height, and transaction index) in the Bitcoin blockchain:

$did : btcr : xyv2 - xzpq - q9wa - p7t$

As aforementioned, creating a DID using did:btcr is achieved by simply creating a Bitcoin transaction. This DID creation transaction may or not refer to a URL that holds a DID document using the OP_RETURN construct. The latter may be stored on a separate storage, e.g., third party server (at the time of writing, IPFS was not supported). In case, the first transaction does not specify an OP_RETURN, a DID document by default is created from the transaction itself. Next operations on the DID (e.g. update transactions) must, however, specify OP_RETURN, otherwise, the DID is considered revoked [17]. An update operation for example, consists of updating the did document and creating new transaction that consumes all previous UTXOs, and that embeds the new link to the updated DID document in the OP_RETURN.

Again, a DID document contains cryptographic material and methods for establishing communication with the DID controller. A verifiable credential issuer (e.g., a university) can then publish their DID, and use it to sign credentials (e.g. diplomas). A verifier (e.g., employer) can check the authenticity of the VC, by resolving the DID that issued the VC, and verifying that it was not revoked using the Bitcoin blockchain. Therefore, the verifier does not have to communicate with the issuer for checking the validity of a given VC, which helps avoid linkability.

3   Bitcoin Privacy Attacks

3.1   Research Method

This section describes the methodology used for collecting and selecting relevant literature, which follows four main steps; (i) research questions identification (cf.

Section 1), (ii) literature search, (iii) literature selection, and (iv) data extraction. **Literature Search.** Collecting relevant literature was carried out through triangulation of a variety of search methods such as manual search and citation search. Scientific databases such as DBLP, IEEE xplore, ACM, usenix and Springer as well as top conferences in the fields of Distributed ledger technology and decentralized identity were searched. **Search Query.** The queries that were employed for searching relevant literature items include and combine the following keywords: "Bitcoin", "blockchain", "distributed ledger technology", "DLT", "privacy", "attack(s)", "anonymity", "deanonymization", "correlation" and "linkability". Only papers from 2009 to 2021 were considered. **Literature Selection.** The search resulted in 479 papers, from which unrelated papers were dropped based on titles and abstracts. Another filtering round based on fast screening of remaining papers resulted in 14 papers that focus on privacy attacks in Bitcoin blockchain. Table 1 lists the venues that were identified ordered by their h5-index and h5-median.

A number of the selected studies have identified privacy attacks, which may reveal links between identities and the Bitcoin addresses. In the following, we categorize and explain the possible attacks that have been applied in the selected papers (Table 2). Based on the paper purposes we categorized the selected papers in five categories including privacy challenges, classification, illicit activities (tracking Bitcoin usage in dark web, ransomware and ponzi schemes), link pseudonyms to IPs, and pattern detection (to find specific patterns related to users behavior in trading systems and remuneration pattern). Some of the papers proposed attacks not specific to Bitcoin. In this paper, we only consider analyzing privacy attacks in the Bitcoin blockchain as we only address Bitcoin privacy attacks in BTCR. Additionally, we employ the four main categories of privacy attacks as identified in [18]; (i) heuristics, (ii) side channel attacks, (iii) flow analysis, and (iv) auxiliary information , which will be explained in next sections.

3.2 Bitcoin Blockchain Heuristics

Table 3 summarizes heuristics that were applied to the Bitcoin protocol to identify relationships between addresses (common input ownership, change address detection, address reuse, single input single output, and specific patterns). One of the heuristics (Cluster growth) prevents false positives [10,21].

Multi/common Input Ownership. The heuristic assumes that the inputs of a transaction are controlled by the same entity and associates all the inputs to one entity. Since the input of a transaction can only be redeemed by providing its signature, it is unlikely that different users join to create a transaction [18]. Figure 2 illustrates the heuristic, where it is assumed that all the addresses (A_1, A_2, A_3) are controlled by one entity (Alice). To prevent false positives, CoinJoin transactions are excluded in the analysis [21]. CoinJoin [29] is one of the most prominent mixing techniques that has been adopted in practice. In mixing techniques, users mix their unspent transaction outputs (UTXOs) with the other users' UTXOs to obfuscate the relationships between the inputs and outputs.

Table 1. Computer security and cryptography top publications

	Publication	h5-index	h5-median	Publisher
1	ACM Symposium on Computer and Communications Security	88	140	ACM
2	IEEE Transactions on Information Forensics and Security	86	118	IEEE
3	USENIX Security Symposium	80	129	USENIX
4	IEEE Symposium on Security and Privacy	74	142	IEEE
5	Network and Distributed System Security Symposium (NDSS)	71	111	NDSS
6	International Conference on Theory and Applications of Cryptographic Techniques (EUROCRYPT)	61	89	SPRINGER
7	Computers & Security	59	90	ELSEVIER
8	IEEE Transactions on Dependable and Secure Computing	54	77	IEEE
9	International Cryptology Conference (CRYPTO)	52	87	SPRINGER
10	International Conference on Financial Cryptography and Data Security	46	74	SPRINGER
11	International Conference on The Theory and Application of Cryptology and Information Security (ASIACRYPT)	42	61	SPRINGER
12	Security and Communication Networks	40	51	Wiley
13	Theory of Cryptography	38	58	SPRINGER
14	ACM on Asia Conference on Computer and Communications Security	37	55	ACM
15	Proceedings on Privacy Enhancing Technologies	35	55	
16	IEEE European Symposium on Security and Privacy	34	74	IEEE
17	Designs, Codes and Cryptography	34	50	SPRINGER
18	European Conference on Research in Computer Security	34	43	SPRINGER
19	IEEE Security & Privacy	31	53	IEEE
20	Journal of Information Security and Applications	31	40	ELSEVIER

In CoinJoin. the users jointly create and sign a transaction to obfuscate the common input ownership heuristic. CoinJoin transactions should be created in the form of the equal-size output to prevent linking the input and output addresses which makes them distinguishable in the blockchain.

Change Address. The heuristic assumes that the change address of a transaction is controlled by the owner of the inputs [10]. The following is a list of common heuristics that are employed to identify change addresses.

Table 2. Selected papers

Category	Paper	Year	Publication	Purpose	Blockchain
Privacy challenges	[19]	2018	IEEE S&P	Access privacy challenges	BTC, ZEC
Classification	[20]	2014	FC	User classification	BTC
	[21]	2020	USENIX	Analysis tool	BTC, BCH, BSV, LTC, and ZEC
Illicit activites	[22]	2018	Computer & Security	Tracking ransomware	
	[23]	2018	IEEE S&P	Tracking ransomware	BTC
	[24]	2019	NDSS	Crypto in dark web	BTC
	[25]	2020	Asia CCS	MMM ponzi detection	BTC
Link Pseudonyms to IPs	[26]	2014	FC	Link Pseudonyms to IPs	BTC
	[27]	2014	CCS	Link Pseudonyms to IPS	BTC
	[28]	2017	FC	Clustering heuristics+network layer info	BTC
	[11]	2019	EuroS&P	Link Pseudonyms to IPs	BTC, ZEC, XMR, Dash
Pattern detection	[12]	2017	EuroS&P	Remuneration detection	BTC
	[13]	2019	CCS	Tracing trading transactions	BTC
	[14]	2019	USENIX	Tracing trading transactions	ETH, BTC, LTC, BCH, Doge, Dash, ETC, ZEC

- *Fresh address:* A fresh address output can be a change address if the other address appeared before in the blockchain [10].
- *Script types:* The only output with a similar script, if all the inputs have similar scripts (e.g., Pay-to-PubkeyHash (P2PKH), Pay-to-Script-Hash(P2SH)) can be a change address [30].
- *Same input and output:* An input address that is also an output address of a transaction can be a change address [30].
- *Optimal change:* An output that has a smaller amount than all the inputs can be a change address [30].
- *Round numbers:* The non-round number output value can be a change address [30,31], since the payment amount is typically a round number.
- *Wallet fingerprinting:* Wallets create transactions in a different manner, which can be used to reveal change addresses [31] (e.g. the change output index, locktime behavior match [30])
- *Peeling chain:* In the peeling chain transactions (transactions where a single address with large amounts pay small amounts to other addresses), the output that continues the peeling can be a change address [30].

Table 3. Bitcoin blockchain heuristics

	Multi-input	Change address	Address-reuse	Single in-single out	Cluster growth	Patterns
[20]	✓	✓				
[19]						
[21]	✓ (excluding CoinJoin)	✓	✓			
[22]	✓	✓				
[23]	✓ (excluding CoinJoin)					
[24]	✓ (excluding CoinJoin)	✓			✓	
[25]						
[26]						
[27]						
[28]	✓	✓			✓	
[11]						
[12]						✓ Remuneration profile
[13]	✓					
[14]	✓			✓		✓ Common relationship

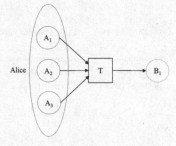

Fig. 2. Multi/common input ownership heuristic

Address Reuse. Whenever the same address is reused, it relates the current transaction to all the transactions that the address previously appeared in. This results in a possible correlation between all transactions enabled with the same address[32]. There is also forced address reuse where the attacker pays a small amount of bitcoin to the used address of the target and follow the address in the blockchain to find other UTXOs belonging to the user if the target combines this UTXO to her other UTXOs in the following transactions [31].

Single Input Single Output. The transaction with only one input and one output is considered as self-payment and the input and the output addresses can be associated with one entity. Indeed, in most cases, the payment transactions consist of multiple inputs and outputs [21].

Cluster Growth. Clusters normally grow in small steps, and if applying a heuristic creates a large cluster, it would be as a result of false positives [28].

Specific Patterns. New heuristics based on the patterns extracted from users and transaction behaviors can be employed. For instance, [12] found remuneration patterns based on the analysis of their ground truth. In [14], they found the common relationship between the addresses in the trading services, where receiving the coins from the same address or sending the coins to the same address can indicate a common social relationship.

3.3 Side Channel Attacks

The correlation of the information such as time, amount, network information, or the user's behavior in the forked blockchains may be used to reveal users or transaction behaviors, thereby compromising their privacy. Table 4 summarizes such side channel attacks, which we explain in the following.

Time Correlation. The attacker can correlate the time that a transaction is confirmed (considering appropriate thresholds) with the time that a user interacted with other services. In the table we provided the research [13,14] that used this attack to find the transactions in the trading services and correlate them with the blockchain data to find the related transactions.

Amount Correlation. The attacker can correlate the amount that has been transferred in blockchain with the amount that has been paid (either by fiat or other crypto currencies) in other services where the latter can be publicly seen via websites or application programming interfaces (APIs). In [13,14], the public trades amounts available in trading services were used to find the corresponding transactions on the blockchain. The attacker can also obtain exchange rates of the fiat currency (if it is paid by fiat currencies) for the date and the time when the transaction is confirmed, and look it up in this interval.

Network Layer Information. The propagation of transactions between nodes can reveal the data in the network layer. The research [11,26–28] indicated the possibility of linking the IP addresses of the nodes to the transactions. To this end, they connected to the Bitcoin nodes and listen to the network to find the original node that is the first who propagated the transaction. It is also mentioned that the access pattern can be used to relate the user to a cryptocurrency address. For instance, visiting a web page with a donation address and then performing a transaction and checking the confirmation in a block explorer can provide an access pattern to link the IP address to that transaction [19].

Cashing Out on Forks. Cross-chain clustering can create a single chain clustering based on the information obtained from the forked chain cluster. This attack links the addresses in one chain based on the activity of those addresses in the forked chain [21]. Researchers [21] combined the Bitcoin and Bitcoin cash clusters and found that the privacy of almost 5% of the Bitcoin transactions is in danger based on their cash-out behaviors in the Bitcoin Cash.

3.4 Flow Analysis

The attacker is able to trace the flow of the money by transaction graph, user graphs, and taint analysis. Table 4 lists the publications that applied the graphs for their analysis.

Transaction Graph. In the transaction graph, the addresses are nodes and the transactions are edges, and the attacker can find predecessors and successors by this graph [20]. Figure 3 illustrates a sample transaction graph where Alice holds 6 BTC, and sends 2 BTC from her address A_1 to Bob B_1 via transaction T_1, as she has 6 BTC in her address, she gets back 4 BTC in her change address A_2. Bob then sends 3 BTC to Carol via transaction T_2 using 2 BTC which he has previously received from the output of T_1 and 2 BTC from the output of another transaction. As can be seen, T_2 has two outputs by which Bob gets 1 BTC as his change (B_3) and pays 3 BTC to Carol.

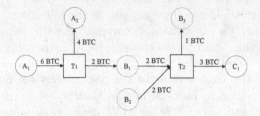

Fig. 3. Transaction graph, adopted from [33]

Taint Analysis. This analysis tracks the flow of the money from an address to another [10]. It is defined as the percentage of the balance of the output address that comes from an input address [18].

User Graph. In the user graph, users are nodes and the transactions are edges which creates the clusters [20] (e.g. by using the heuristics), this graph can find the relationship between different users in the blockchain.

Table 4. Side channel attacks and flow analysis

	Time correlation	Amount correlation	Network layer		Cashing out on forks	TX graph/ User graph	Taint analysis
			Map IP to pseudonyms	Access pattern/ user behavior pattern			
[20]						✓	
[19]			✓	✓			
[21]					✓ (Combining BCH& BTC)	✓	
[22]						✓	
[23]	✓					✓	
[24]						✓	✓
[25]							
[26]			✓				
[27]			✓				
[28]			✓			✓	
[11]			✓				
[12]		✓ (BTC in $)					
[13]	✓	✓				✓	
[14]	✓	✓					

3.5 Auxiliary Information

The attacker can tag the addresses using several ways including searching on the Internet, interacting with the target, using service APIs, etc. The aforementioned heuristics, side-channel attacks, and flow analysis find the relationship between the addresses, therefore, if the attacker tag an address, he is able to tag other addresses related to this address. The attacker can not only tag the addresses but also in some cases, can obtain information about the locations, emails, usernames, and etc. Table 5 indicates the resources which have been used in the selected papers to tag the addresses. Some entities publish their addresses in forums, social networks, and websites. As can be seen in table 5, Bitcointalk, Reddit, Twitter are well-known resources to find Bitcoin addresses [13,20,22]. Addresses published on the Websites, and addresses which can be queried in the search engines can identify the information about addresses. Services' APIs also provide some additional information that can be related to the addresses (e.g. the information from trading services, such as Localbitcoins, Changelly, Shapeshift) [13,14,20]. Some attackers interact with services to obtain the addresses belonging to a specific service [23], it is also called mystery shopper payment [31] where the attacker pays a small amount and follows the address associated with the service in the blockchain. They are non-commercial and commercial databases that provide the address tags based on the ground truth, they found. Walletexplorer, Chainalysis, blockchain.info, and some of the researchers who published their address tags are examples of such databases that were used to tag the addresses [14,23,25]. In table 5 the resources that the previous research utilized to tag the addresses are provided.

Table 5. Auxiliary information resources

	Forums	Websites	Search engines	Social networks	Service APIs	Interacting	Address tags DB	others
[20]	✓BitcoinTalk, Bitcoin-OTC	✓Casascius physical coins	✓Google	✓Reddit	✓Mt.Gox		✓blockchain.info	
[19]								
[21]								
[22]		✓BleepingComputer, MalwareTips, 2-spyware	✓Google, Yahoo	✓Reddit				✓†
[23]	✓BleepingComputer±±	✓ID ransomware				✓**	✓Chainalysis	✓Synthetic addr ††
[24]			✓Ahmia, FreshOnions, Google				✓Walletexplorer,	
[25]	✓BitcoinTalk *						✓Walletexplorer, blockchain.info	
[26]								
[27]								
[28]								
[11]								
[12]								
[13]	✓BitcoinTalk	✓Localbitcoins		✓Twitter	✓Localbitcoins		✓Walletexplorer,	
[14]					✓Changelly, Shapeshift		✓Walletexplorer, researchers data,	

±±Ransom addresses in Bleeping computer forum.
† Ransomware knowledge base, YouTube videos, reports from Counter Threat Units (CTU), Incident Responses (IR), and Security Operations Centers (SOC).
†† By running ransomware binaries.
* Extracting Ponzi addresses, profile information, age, gender, location, ...
** Paying a small amount to ransom addresses.

4   BTCR Privacy Issues and Possible Countermeasures

In this section, we investigate the privacy of the method did:btcr based on the adopted criteria from RFC 6973 [9] including surveillance, misattribution, correlation, identification, secondary use, and disclosure.

4.1   Surveillance

Any kind of observation and monitoring of the users is considered as surveillance, whether the users are aware of the surveillance or not, it can influence the privacy of the user [9]. In the previous section, we showed the possibility of monitoring users and linking the real-world identities to the Bitcoin addresses which tends to compromise users' activities and their economic situations. The surveillance of DID in the Bitcoin blockchain can be investigated in different aspects. The auxiliary information can be obtained through the interactions with services using DIDs. The service is, therefore, able to follow the user's activities and money flow in the blockchain using Bitcoin privacy attacks. For instance, a payment service where a user authenticates to the device using a DID and then pays using another Bitcoin address that belongs to her, associates the DID to that Bitcoin address. Furthermore, the privacy concerns that a user should take into account when using an immutable blockchain for creating DIDs have a significant role. A user who is aware of such problems can employ privacy-preserving techniques to protect herself against such privacy attacks. A previous research [34,35] indicated a misconception in the privacy of the Bitcoin blockchain, which can result in serious problems for applications that use blockchain technology. Using Tor services [15], mixing the UTXO before using it for DID BTCR [8] to unlink the relationship between the BTCR UTXO and other UTXOs belonging to the user, and to prevent combining the revoked DID BTCR with other UTXOs in the future to spend the amount associated with the UTXO can be used as possible countermeasures to surveillance of the DID BTCR.

4.2   Misattribution

Misattriubation is considered whenever a user's data or communications are attributed to another, which can consequently affect the user's reputation [9]. Some of the indistinguishable mixing techniques such as PayJoin [36] can relate the users' UTXOs to someone else, using the common input ownership heuristic [37]. PayJoin [36] is one of the successors of the CoinJoin technique, where a user creates a CoinJoin transaction by the recipient of the transaction. The recipient adds her coins as an input of the transaction which consequently increases the payment amount. Therefore, this technique does not require an equal-size output and it is indistinguishable in the blockchain. This would cause privacy problems for the users who are not aware of this issue when using PayJoin as a privacy technique or interacting with the service that implemented PayJoin (e.g., merchants, exchange). Therefore, using the Bitcoin blockchain for DIDs in did:btcr can put the users at the risk of such privacy misattribution. Providing information for the users to inform them from the possible misattribution by using specific mixing techniques such as PayJoin for the UTXOs that are used in BTCR can to some extent prevent this privacy problem.

4.3 Correlation

Correlation is considered as the combination of different information, which relate to one user [9]. We discuss the correlation in three different aspects including DIDs and DID documents correlation, time correlation, and network correlation. (i) Using the same DID or DID document for interacting with different services can help to trace and correlate user activities [5,15,38]. Furthermore, using the same public keys in different DID documents can reveal the link between the corresponding DIDs, e.g., interactions with different services using the same DID while showing different VCs. Inversely, if different DIDs are used for each service while using the same DiD document, then those services can associate those multiple DIDs to the same user. Pairwise-unique DIDs that are issued on a per-relationship basis which can not be correlated to each other or single-use identifier that is discarded once it is exchanged can be used to mitigate this issue [39]. Another issue is that the DID document contains methods for verification of the DID and the attributes including "also known as" and "controller" [15]. Using "also known as", it is possible to specify another identifier belonging to the same user. This can be useful for businesses that use multiple DIDs for their services but should be avoided if not required. Using "controller", another entity can be specified, which is then allowed to change the DID document or to authenticate. This may reveal a relationship between the subject and the controller DiD if they are different. (ii) Considering the network layer correlation, the IP address of the entity can compromise the relationship of common controls, where an attacker can identify the link between different DIDs based on the IP address of the clients [39]. Additionally, using traffic analysis by checking the access history to the DID documents, may help correlate IP addresses to the DID documents. Using TOR or proxy can provide additional privacy [15] in this regard. (iii) Time correlation by employing the same service endpoints can be used to find the relationship of common controls [39]. For instance, timing analysis can be used to correlate users' activities whenever a user uses the same service endpoint in the DID documents. Sharing the service endpoints between a variety of DIDs that are controlled by the different entities [15] can be considered as a possible countermeasure.

4.4 Identification

Identification is considered as relating the information to a specific user to derive her identity [9]. Storing any type of personally identifiable information (PII) in the blockchain, even encrypted or hashed, has the potential to put the users' privacy at risk, as they may be broken and be publicly accessible [5,7,15,38]. Despite DID revocation support, the immutability property prevents deleting the logs of existing BTCR DIDs. Therefore, if the Bitcoin address associated to a DID is later spent with some other inputs without using mixing techniques (will also be considered as revoked), it can link the address used for DID to other addresses owned by the user, based on the common input ownership heuristic. Moreover, if a transaction in the BTCR (when it is revoked) contains a change address, it can be linked to the owner of the inputs. Thus, it is suggested to create the transactions without a change address. Not only Blockchain analysis can identify real-world identities and relate them to DIDs, but also metadata tracing in the DID documents can provide information in the identification of the entities. The visibility of the DID document can leak the metadata about the attributes [6] and provide information about the service endpoints. In BTCR, the attacker can query the Bitcoin blockchain to identify all transactions with OP_RETURN that specify a link to a DID document, thus enabling access to metadata and associated service endpoints. To prevent any privacy leak, URLs to the service endpoints should not include any personal information (e.g. usernames). Usually, the DID documents are stored on servers. If the DID document is stored in the third-party server, the latter may identify the real DiD owner. If the DiD document is stored on a user own server, it becomes possible to correlate the user IP address with the DID document. In this case IPFS (The InterPlanetary File System)[3] can be used as a countermeasure.

4.5   Secondary Use

Secondary use is considered as collecting the information about a user without her consent and using it for different purposes other than which the information was collected [9]. We investigate secondary use in did:btcr in three aspects. (i) Read/resolve makes it possible to trace the DID use if it is accessed by third-party services (e.g., universal DID resolver, a naive implementation of Simplified Payment Verification (SPV) clients [40], checking the DID on block explorers), in this case, the attacker can find the resolution pattern. To prevent third party services from collecting information about users, the latter may employ their own Bitcoin full nodes. (ii) The verifier is able to trace the transaction flow, check the history of the UTXOs (e.g. user activities), and if they are spent (accidentally or for changing the ownership, or revocation) monitor next transactions' flow. The verifier can also see all the amounts associated with the address. (iii) a DID real identity can be compromised if used in services that require information about them or their activities (e.g. social networks).

4.6   Disclosure

Disclosure is considered as exposure of information about a user which violates the confidentiality of the shared data [9]. All the privacy attacks that were mentioned in the previous sections can be applied to the addresses that are used as DIDs in did:btcr. The users who are not familiar with the privacy issues in the Bitcoin blockchain may encounter some serious problem if their DIDs' addresses

---

[3]   https://ipfs.io/.

link to their other addresses in the blockchain. This tends to lose privacy in their economic activities for the services that they are authenticated by DIDs. To create the first DID in BTCR, the user should provide an address, where she can buy from an exchange. The latter has access to information related to the owner (email address, etc.) or in some cases the real identity of the owner when KYC (know your customer) is applied. The user can use mixing techniques [8] beforehand to obfuscate the relationship between the UTXO used in BTCR and the other UTXOs beloging to her. The BTCR updates are required to include the OP_RETURN field; therefore, the users can not utilize current mixing techniques to provide better privacy for their associated addresses. This makes the BTCR updates traceable in the Bitcoin blockchain. Thus, every update in BTCR not only reveals the public key of the previous DID but also indicates the update or changing the access control.

5   Conclusion

In this paper, we presented a review of Bitcoin privacy attacks, which we categorized into four main categories. Then, we investigated and analyzed six possible privacy threats to the DID method did:btcr. In particular, we showed how data analysis of Bitcoin public records, in combination with auxiliary information can be exploited using sophisticated heuristics, to reveal or correlate transactions, identities, or addresses of users. This information, in turn, may be used by malicious actors and cybercriminals to conduct, for example, extortion or ransomware attacks. This study has demonstrated that although BTCR provides some advantages such as protection against censorship, integrity, access and a degree of decentralization, it still lacks methods to deal with the privacy issues identified in this paper. Future research will consist on elaborating and developing new methods, or using existing privacy-enhancing techniques (e.g., mixing techniques, zero-knowledge proofs) to address the aforementioned privacy issues.

Acknowledgments. This research is based upon work partially supported by (1) SBA Research (SBA-K1); SBA Research is a COMET Center within the COMET – Competence Centers for Excellent Technologies Programme and funded by BMK, BMDW, and the federal state of Vienna. The COMET Programme is managed by FFG. (2) the FFG ICT of the Future project 874019 dIdentity & dApps. (3) the FFG Basisprogramm Kleinprojekt 39019756 Decentralised Marketplace for Digital Identity.

References

1. Nakamoto, S.: Bitcoin: a peer-to-peer electronic cash system. In: Decentralized Business Review, p. 21260 (2008)
2. López-Pintado, O., García-Bañuelos, L., Dumas, M., Weber, I., Ponomarev, A.: CATERPILLAR: a business process execution engine on the ethereum blockchain. CoRR abs/1808.03517 (2018)

3. Ladleif, J., Weber, I., Weske, M.: External data monitoring using oracles in blockchain-based process execution. In: Asatiani, A., et al. (eds.) BPM 2020. LNBIP, vol. 393, pp. 67–81. Springer, Cham (2020). https://doi.org/10.1007/978-3-030-58779-6_5

4. Prybila, C., Schulte, S., Hochreiner, C., Weber, I.: Runtime verification for business processes utilizing the bitcoin blockchain. Future Gener. Comput. Syst. **107**, 816–831 (2020)

5. Lesavre, L., Varin, P., Mell, P., Davidson, M., Shook, J.: A taxonomic approach to understanding emerging blockchain identity management systems. arXiv preprint arXiv:1908.00929 (2019)

6. Dunphy, P., Petitcolas, F.A.: A first look at identity management schemes on the blockchain. IEEE Secur. Privacy **16**(4), 20–29 (2018)

7. Allen, C., Hamilton Duffy, K., Grant, R., Pape, D.: BTCR did method. https://w3c-ccg.github.io/didm-btcr/ (2019)

8. Ghesmati, S., Fdhila, W., Weippl, E.: Bitcoin privacy - a survey on mixing techniques. Cryptology ePrint Archive, Report 2021/629 (2021). https://eprint.iacr.org/2021/629

9. Cooper, A., et al.: Privacy considerations for internet protocols. Internet Architecture Board (2013)

10. Meiklejohn, S., et al.: A fistful of bitcoins: characterizing payments among men with no names. In: Proceedings of the 2013 Conference on Internet Measurement Conference, pp. 127–140 (2013)

11. Biryukov, A., Tikhomirov, S.: Deanonymization and linkability of cryptocurrency transactions based on network analysis. In: IEEE European Symposium on Security and Privacy (EuroS&P), vol. 2019, pp. 172–184. IEEE (2019)

12. English, S.M., Nezhadian, E.: Conditions of full disclosure: The blockchain remuneration model. In: IEEE European Symposium on Security and Privacy Workshops (EuroS&PW), vol. 2017, pp. 64–67. IEEE (2017)

13. Sabry, F., Labda, W., Erbad, A., Al Jawaheri, H., Malluhi, Q.: Anonymity and privacy in bitcoin escrow trades. In: Proceedings of the 18th ACM Workshop on Privacy in the Electronic Society, pp. 211–220 (2019)

14. Yousaf, H., Kappos, G., Meiklejohn, S.: Tracing transactions across cryptocurrency ledgers. In: 28th {USENIX} Security Symposium ({USENIX} Security 19), pp. 837–850 (2019)

15. Reed, D., Sporny, M., Longley, D., Allen, C., Grant, R., Sabadello, M., Holt, J.: Decentralized identifiers (dids) v1. 0. Draft Community Group Report (2021)

16. Sporny, M., Noble, G., Longley, D., Burnett, D., Zundel, B.: Verifiable credentials data model (2019)

17. Wiki: Op_return. https://en.bitcoin.it/wiki/OP_RETURN (2020)

18. Bonneau, J., Narayanan, A., Miller, A., Clark, J., Kroll, J.A., Felten, E.W.: Mixcoin: anonymity for bitcoin with accountable mixes. In: Christin, N., Safavi-Naini, R. (eds.) FC 2014. LNCS, vol. 8437, pp. 486–504. Springer, Heidelberg (2014). https://doi.org/10.1007/978-3-662-45472-5_31

19. Henry, R., Herzberg, A., Kate, A.: Blockchain access privacy: challenges and directions. IEEE Secur. Privacy **16**(4), 38–45 (2018)

20. Spagnuolo, M., Maggi, F., Zanero, S.: BitIodine: extracting intelligence from the bitcoin network. In: Christin, N., Safavi-Naini, R. (eds.) FC 2014. LNCS, vol. 8437, pp. 457–468. Springer, Heidelberg (2014). https://doi.org/10.1007/978-3-662-45472-5_29

21. Kalodner, H., et al.: Blocksci: design and applications of a blockchain analysis platform. In: 29th {USENIX} Security Symposium), pp. 2721–2738 (2020)

22. Conti, M., Gangwal, A., Ruj, S.: On the economic significance of ransomware campaigns: a bitcoin transactions perspective. Comput. Secur. **79**, 162–189 (2018)

23. Huang, D.Y., et al.: Tracking ransomware end-to-end. In: IEEE Symposium on Security and Privacy (SP), vol. 2018, pp. 618–631. IEEE (2018)

24. Lee, S., et al.: Cybercriminal minds: an investigative study of cryptocurrency abuses in the dark web. In: NDSS (2019)

25. Boshmaf, Y., Elvitigala, C., Al Jawaheri, H., Wijesekera, P., Al Sabah, M.: Investigating MMM Ponzi scheme on bitcoin. In: Proceedings of the 15th ACM Asia Conference on Computer and Communications Security, pp. 519–530 (2020)

26. Koshy, P., Koshy, D., McDaniel, P.: An analysis of anonymity in bitcoin using P2P network traffic. In: Christin, N., Safavi-Naini, R. (eds.) FC 2014. LNCS, vol. 8437, pp. 469–485. Springer, Heidelberg (2014). https://doi.org/10.1007/978-3-662-45472-5_30

27. Biryukov, A., Khovratovich, D., Pustogarov, I.: Deanonymisation of clients in bitcoin p2p network. In: Proceedings of the 2014 ACM SIGSAC Conference on Computer and Communications Security, pp. 15–29 (2014)

28. Neudecker, T., Hartenstein, H.: Could network information facilitate address clustering in bitcoin? In: Brenner, M., et al. (eds.) FC 2017. LNCS, vol. 10323, pp. 155–169. Springer, Cham (2017). https://doi.org/10.1007/978-3-319-70278-0_9

29. Maxwell, G.: Coinjoin: Bitcoin privacy for the real world (2013). https://bitcointalk.org/index.php

30. Kalodner, H.: Privacy. https://citp.github.io/BlockSci/reference/heuristics/change.html. Accessed 23 July 2020

31. Wiki: Privacy. https://en.bitcoin.it/wiki/Privacy. Accessed 23 July 2020

32. Wiki: Address reuse (2021). https://en.bitcoin.it/wiki/Address_reuse

33. Möser, M., Böhme, R., Breuker, D.: An inquiry into money laundering tools in the bitcoin ecosystem. In: APWG eCrime Researchers Summit, vol. 2013, pp. 1–14. IEEE (2013)

34. Mai, A., Pfeffer, K., Gusenbauer, M., Weippl, E., Krombholz, K.: User mental models of cryptocurrency systems–a grounded theory approach (2020)

35. Krombholz, K., Judmayer, A., Gusenbauer, M., Weippl, E.: The other side of the coin: user experiences with bitcoin security and privacy. In: Grossklags, J., Preneel, B. (eds.) FC 2016. LNCS, vol. 9603, pp. 555–580. Springer, Heidelberg (2017). https://doi.org/10.1007/978-3-662-54970-4_33

36. Gibson, A.: Payjoin (2018). https://joinmarket.me/blog/blog/payjoin/

37. Ghesmati, S., Kern, A., Judmayer, A., Stifter, N., Weippl, E.: Unnecessary input heuristics and PayJoin transactions. In: Stephanidis, C., Antona, M., Ntoa, S. (eds.) HCII 2021. CCIS, vol. 1420, pp. 416–424. Springer, Cham (2021). https://doi.org/10.1007/978-3-030-78642-7_56

38. (W3C), C.C.G.: A primer for decentralized identifiers (2020). https://w3c-ccg.github.io/did-primer/

39. Andrieu, J., et al.: Did method rubric v1.0 (2021). https://w3c.github.io/did-rubric/#privacy

40. Wiki: Simplified payment verification (2019). https://en.bitcoinwiki.org/wiki/Simplified_Payment_Verification

Enhancing Blockchain-Based Processes with Decentralized Oracles

Davide Basile[1] ⓘ, Valerio Goretti[1] ⓘ, Claudio Di Ciccio[1(✉)] ⓘ,
and Sabrina Kirrane[2] ⓘ

[1] Sapienza University of Rome, Rome, Italy
{basile.1810355,goretti.1811110}@studenti.uniroma1.it,
claudio.diciccio@uniroma1.it
[2] Vienna University of Economics and Business, Vienna, Austria
sabrina.kirrane@wu.ac.at

Abstract. The automation of business processes via blockchain-based systems allows for trust, reliability and accountability of execution. The link that connects modules that operate within the on-chain sphere and the off-chain world is key as processes often involve the handling of physical entities and external services. The components that create that link are named oracles. Numerous studies on oracles and their implementations are arising in the literature. Nevertheless, their availability, integrity and trust could be undermined if centralized architectures are adopted, as taking over an oracle could produce the effect of a supply-chain attack on the whole system. Solutions are emerging that overcome this issue by turning the architecture underneath the oracles into a distributed one. In this paper, we investigate the design and application of oracles, distinguishing their adoption for the in-flow or out-flow of information and according to the initiator of the exchange (hence, pull- or push-based).

Keywords: Decentralized applications · Business process management · Distributed architectures

1 Introduction

Since its inception, the technologies related to the blockchain world are constantly evolving. In particular, its decentralized aspect has offered a development environment for Decentralized Applications (DApp), where data integrity and consistency are crucial factors [12]. However, applications developed on such platforms are unable to obtain information from the off-chain world, and cannot directly alter the outer world status [3,5]. Therefore, intermediate components named oracles have been introduced to open up the blockchain to the real world [18,25].

One of the usages in which DApps have shown potential is the coordination of business processes between multiple parties [8,16]. Especially in this scenario, oracles represent the trusted link with external sources of information. The possibility of erroneous or counterfeit information can result in major financial implications for the various stakeholders [13,23]. By distributing and decentralizing the

© Springer Nature Switzerland AG 2021
J. González Enríquez et al. (Eds.): BPM 2021, LNBIP 428, pp. 102–118, 2021.
https://doi.org/10.1007/978-3-030-85867-4_8

transmitted information, and using redundancy procedures, the likelihood of such problems is greatly reduced – which is also one of the driving factors at the core of blockchains. Indeed, decentralization increases the robustness and security of transmission operations by removing the problems associated with a single point of failure [11].

This paper studies the effect of decentralizing blockchain oracle architectures in terms of availability, integrity and trust. In particular, we examine the design and the implementation of decentralized and centralized oracles for the Ethereum platform, categorized as per the patters described in [18]. Unlike the typical scenario for oracles, we consider cases in which different off-chain sources retain separate parts of an information to be collected, or separate targets receive information from the blockchain. The proposed implementations are then evaluated in terms of both latency and costs.

The remainder of the paper is structured as follows. Section 2 presents the necessary background in terms of blockain platforms, the Ethereum ecosystem and the role of blockchain oracles. Section 3 introduces the motivating use case scenario used to guide our work. Section 4 sketches our blockchain oracle conceptual framework. Section 5 provides and overview of our performance evaluation, while Sect. 6 identifies open challenges and opportunities. Finally, we present our conclusions and plans for future work in Sect. 7.

2   Background

In the following, we briefly present background information on blockchains, with a special focus on the Ethereum ecosystem, and blockchain oracles.

2.1   Blockchain: Definition and Applications

A blockchain is a protocol for the distributed management of a data structure in which transaction are stored sequentially in an append-only list (the ledger). Updates on the ledger are communicated via sequential blocks that are built and validated (i.e., mined), and then broadcasted among the nodes in the network. The ledger is replicated in all nodes of the network. Nodes agree on the inclusion of the next block information via consensus algorithms [26]. Its decentralized, persistent and immutable characteristics make blockchain suitable for the needs of automated systems in which interactions between multiple untrusted parties are recorded [10]. Such systems have long been primarily used for payments via cryptocurrency transactions, as their infrastructure allows for the storage and regulation of exchanges without the arbitration of external authoritative entities [19].

With the advent of Ethereum [5], second-generation blockchain platforms emerged as the blockchain turned from being mainly an e-cash distributed management system to a distributed programming platform at the basis of Decentralized Applications (DApps) [17]. In particular, Ethereum enabled the deployment and run of smart contracts (i.e., stateful software artefacts exposing variables and callable methods) in the blockchain environment through the Ethereum

Table 1. Classification of oracles [18]

Inform. flow direction		Information exchange initiator	
		Pull	Push
	In	An on-chain component requires information from the outside word	An off-chain component initializes the procedure and sends data from outside the blockchain
	Out	An off-chain component requires information from the blockchain	An on-chain component initializes the procedure and sends data from the blockchain to the outside world

Virtual Machine (EVM). The code of the deployed smart contract is stored into the blockchain itself. Every time a user interacts with a smart contract method, a new transaction is generated. As the code is executed by the EVM and not locally, users are required to pay fees (the so-called "gas") as a compensation for the computational power used. The users can specify the maximum limit they would pay and the price per gas unit in terms of the Ethereum native cryptocurrency (Ether). The amount of gas to be paid is proportional to the complexity of the code and the operations involved.

Notice that smart contracts can be invoked from the off-chain and, during the method execution, exchange messages within the on-chain spheres with other smart contracts. However, on-chain code cannot directly invoke off-chain programs for the sake of consistency and determinism. Ethereum smart contracts can emit so-called events [7], namely developer-specified data fields included within transactions that typically mark relevant stages of the execution. Off-chain software artefacts can subscribe to such events to react to the signalled statuses.

The new capabilities unlocked an array of new application domains for blockchains, including sectors like insurance and music and areas such as the internet of things and cybersecurity [1,20]. As emphasized by Tareq Ahram et al. [23], a key application domain for blockchains is supply chain management. In this scenario, blockchains are used to record the data generated in every step of the supply chain, by creating an immutable history of the good produced or the service delivered. In this way, the blockchain can greatly facilitate the recording of assets and the tracking of invoices, payments and orders. The motivating use case scenario in Sect. 3 is rooted in this domain.

2.2 Blockchain Oracles

The variety of DApps developed on the Ethereum blockchain has underlined the need to ensure the robustness, consistency and persistence of blockchain data by defining a structural context in which this technology is proposed as a closed and self-contained system unable to communicate with the outside world [17,25]. The inability of smart contracts to access data that are not already stored on-chain can be a limiting factor for many application scenarios such as that of multi-party business processes. The solution to this problem comes in the form of oracles [24].

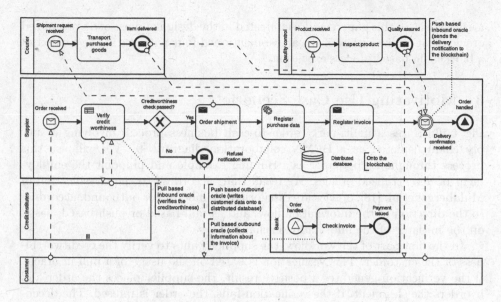

Fig. 1. BPMN diagram of the use case scenario

An oracle can be seen as a bridge that allows for the communication between the on-chain and the off-chain world. The DApp should be able to trust the oracle in the same way as it does so with the information from within the blockchain. Reliability for oracles is key [2,15]. Moreover, an oracle has the arduous task of acting as a link between the blockchain application and different external entities, which may potentially be characterized by different technologies and mechanisms. Therefore, the designation and sharing of a well-defined protocol becomes fundamental for the proper functioning of the service. Mühlberger et al. [18] describe oracle patterns that consider two dimensions: the information direction (inbound or outbound) and the initiator of the information exchange (pull- or push-based). Inbound oracles inject data into the blockchain from the outside, whereas outbound oracles transmit information from the blockchain to the outside. Pull-based oracles are such that the initiator is the recipient of the information, whereas with push-based oracles the initiator is the sender of the information. By combining the push-/pull-based and inbound/outbound classifications, they define four oracle design patterns. Table 1 summarizes these types of oracles: The pull-based inbound oracle (henceforth, *pull-in* oracle for simplicity) is used when an on-chain component starts the procedure and injects data from the real world. The push-based inbound (*push-in* for short) oracle is used by an off-chain component to send data to the blockchain. The pull-based outbound (*pull-out*) oracle is used when an off-chain component needs to retrieve data from the blockchain. Finally, the push-based outbound (*push-out*) oracle allows an on-chain component to transmit information outside the blockchain. In addition to the information direction, Beniiche [3] categorizes existing oracle solutions according to the source of information (human, software or hardware) and on their centralized or decentralized architecture. In this paper, we are interested in the design and use of decentralized oracles that realize either of the above-mentioned four patterns in the context of a blockchain-based process execution.

3   Motivating Use Case Scenario

Fig. 1 illustrates a multi-party order-to-cash business process involving a supply chain depicted as a BPMN collaboration diagram [9]. We will use this process throughout the paper as a running example and pinpoint the employment of decentralized oracles. We recall that, according to the classification of Mühlberger et al. [18], oracles are categorized as inbound or outbound, according to the direction of the information flow, and as pull-based or push-based, based on the initiator of the information exchange.

In the first part of the workflow, the supplier wants to verify the creditworthiness of the customer. This verification is based on the usage of a pull-in oracle. If the verification generates a positive result, the supplier places the order and it orders the shipment. If the verification fails, the order is refused. The decentralized architecture of the oracle allows for the retrieval of distributed information about the creditworthiness, as the customer has open accounts in multiple credit institutions. We assume the sensitive information about the customer to be properly protected from malicious treatment or leakage through the usage of existing privacy-preserving record-linkage techniques [21]. Once the order is placed, and the product is handed to the courier, the shipment procedure starts. In the meantime, the supplier records the data of the purchase order into an external distributed database via a push-out oracle. The decentralized architecture of the oracle fits with the need to send data to a destination consisting of multiple nodes as the distributed database, in our scenario, resorts to physical instances. After that, the supplier registers the invoice in the blockchain. Meanwhile, the courier delivers the ordered product. At the customer's side, a quality control specialist checks that the consigned goods conform with the standards. If so, a push-in oracle uses the blockchain as a notarization means to record that the delivery succeeded. Notice that this passage requires three actors to give their confirmation based on three distinct information bits: the courier (for the consignment), the quality control specialist (for the status of the consigned material) and the customer (for the receipt of the goods). The push-in oracle is thus decentralized as well, as it requires a confirmation from multiple parties. Finally, the banking system can unlock the payment. In response to the notification of the finalized handling of the order, the bank verifies that the invoice is stored on the blockchain. As a successive layer of security, it retrieves the data from multiple, physically distinct blockchain nodes – thereby employing a decentralized pull-based oracle.

In the following section, we show a possible reference architecture for the decentralized version of the aforementioned oracle categories.

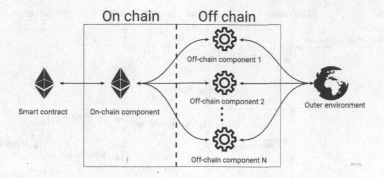

Fig. 2. Decentralized oracles architecture at large

4   Decentralized Oracles

The main limitation that characterizes centralized oracles is the presence of a unique operative unit that works in order to make the information flow between the blockchain and the outer environment. This particular aspect can cause several critical issues that put the entire production chain at risk. The first one is the problem of possibly having a single point of failure as the oracle could be the weak link for cyber attacks. It is interesting to notice that attacks in which a trusted software component is injected with malicious code fall under the name of (software) supply-chain attacks [22]. Indeed, since the architecture provides only one operative unit, a potential malfunction determines the end of (trusted) communication and the potential loss of availability. Moreover, the centralization requires a greater guarantee of correctness of the transmitted information. The single component responsible for the communication cannot tolerate wrong or incoherent data.

Decentralized oracles, instead, resort to multiple independent components that send and receive information. In this case, the oracle becomes an information distribution network regulated by internal protocols such as an own consensus mechanism or incentivization strategies [14]. It is possible to adopt different approaches to the question of oracles' consensus. For example, some systems already on the market such as Gnosis[1] and Augur[2] adopt a voting mechanism combined with human oracles. Other systems, such as Chainlink,[3] propose a fully automated majority approach. The details on the management of those networks go beyond the scope of the paper. We refer to [4] for an overview of the mechanisms underlying an envisioned decentralized oracle network.

Without loss of generality, we assume here a fair behavior of the oracles and an inner consensus algorithm based on the agreement of the totality of the involved components. In the remainder, we will show how decentralization can be applied to oracles.

---

[1] https://www.gnosis.io/ Accessed: July 14, 2021.
[2] https://augur.net/ Accessed: July 14, 2021.
[3] https://chain.link/ Accessed: July 14, 2021.

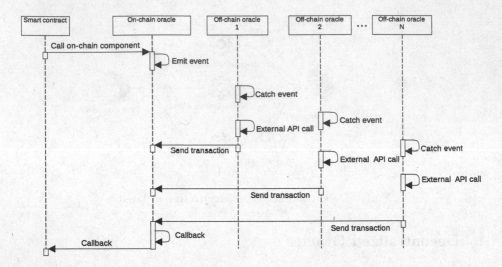

Fig. 3. Sequence diagram of the interactions with a decentralized pull-in oracle

4.1 Architecture Overview

Regardless of the type under consideration, oracles can be split into two main tiers, as illustrated in Fig. 2. The on-chain tier manages the interaction between the oracle system and the on-chain world. It has a single software component inside, namely a full-fledged smart contract that can be seen as an entry point for decentralized applications that want to use that specific oracle system. The ways the DApp interacts with the on-chain tier is defined by the interaction protocol of the oracle itself. The off-chain tier is used to manage the interaction between the real world and the oracle. The two components of the oracle are able to communicate by sending data to each other. In the Ethereum ecosystem, when the on-chain tier sends data to the off-chain tier, it generates a new event containing the relevant information, which is caught in the off-chain tier. The off-chain tier can send data to the on-chain tier by using the methods exposed by the smart contract of the latter via a transaction with the necessary input. The decentralisation of the architecture takes place inside the off-chain tier, as multiple external modules interact with the on-chain world and operate independently, in order to retrieve data from the off-chain environment or send data to it.

In the following, we detail the design of our decentralized oracle architecture for each of the four categories described in [18]. We will refer to the oracles' on-chain components as on-chain oracles and to the components inside the off-chain tier as off-chain oracles for the sake of brevity.

4.2 Decentralized Pull-in Oracle

In a pull-in oracle, the interaction begins with the call from the smart contract implementing the process logic to the on-chain oracle, as depicted in the sequence diagram in Fig. 3. Considering the running example of Sect. 3, the purpose of the pull-in oracle is to connect the decentralized application with multiple credit institutions, in order to verify the creditworthiness of the customer – which is confirmed only if all credit institutes agree. The smart contract running the check activity interacts with the on-chain component of the oracle, generating a new request for verification. The on-chain component, then, emits a new event containing the data to be processed by the off-chain oracle (e.g., the customer personal information). At that point, the off-chain oracles catch the emission of the event, and they execute their business logic (e.g., the creditworthiness verification) based on different data sources via dedicated API calls (the credit institutions). Once the off-chain oracles have obtained the result of their computation, they invoke the on-chain oracle callback method to return the answer via transactions. In our simplified consensus mechanism, we assume that when all the off-chain components have sent their answer to the on-chain smart contract, it uses the callback method of the decentralized application, in turn, to return the aggregate result.

4.3   Decentralized Push-out Oracle

Figure 4 illustrates the interactions that realize the information exchange via a decentralized push-out oracle. Unlike the pull-in oracle, the source of information lies within the blockchain, which is by its nature a decentralized system. The procedure for pushing out the information starts when the smart contract creates a new request for the outbound transfer of data. The on-chain oracle generates a new event that contains the data to be exposed. When the off-chain components catch the event, they all operate independently, invoking external APIs. In our scenario, the purchase order data are stored in an external distributed database. In order to update the database with new orders, the smart contract

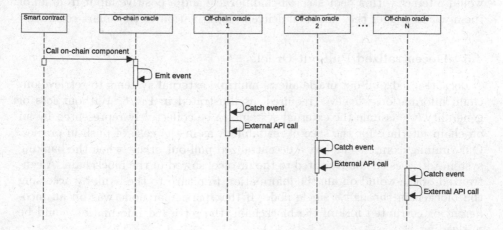

Fig. 4. Sequence diagram of the interactions with a decentralized push-out oracle

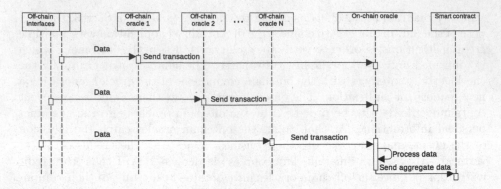

Fig. 5. Sequence diagram of the interactions with a decentralized push-in oracle

underpinning the activity execution employs a push-out oracle and every off-chain oracle interacts with a different instance of the distributed database.

4.4 Decentralized Push-in Oracle

Figure 5 depicts the transfer of information into a blockchain mediated by a decentralized push-in oracle. We assume an off-chain interface gathers data from various sources and sends it to the off-chain oracles. In turn, the off-chain oracles send the transaction with those data to the on-chain oracle, which is responsible for the collection of the different pieces of information, their aggregation and final communication with the smart contract. In our scenario, a decentralized push-in oracle is used to confirm that the delivery was successful, upon the notification from three different off-chain information providers, namely the carrier, the customer and the quality control specialist. Each of those information providers would interact with a dedicated off-chain oracle and a positive input from all of them would trigger the successful delivery confirmation to the smart contract.

4.5 Decentralized Pull-out Oracle

A decentralized pull-out oracle allows multiple external systems to retrieve on-chain information whenever required, as illustrated in Fig. 6. Without loss of generality, we assume the external systems to be collectively represented by an off-chain interface for the sake of readability, as in the case of push-in oracles. Our running example employs a decentralized pull-out oracle when the banking system retrieves the data related to the invoice, stored in the blockchain. A centralized oracle would obtain the information to return to the bank by accessing the blockchain through a single node. If that particular node was on an incoherent or corrupted fork of the blockchain, the retrieved information could be misleading.

The decentralized version of the oracle is used here to overcome the potential inconsistency of the blockchain data through its own decentralized nature, i.e., by resorting to several independent components that watch the blockchain via different nodes. The process starts when the off-chain interface (invoked, e.g., by the banking system) requests data (e.g., the invoice information) to multiple off-chain oracles. Each of them generates a new query towards different nodes of the blockchain. In every node, the on-chain oracle would return the current response based on the local view of the blockchain, in turn given back to the requesting off-chain interface.

5   Implementation

In this section, we briefly describe a proof-of-concept prototype implementing the decentralized oracle architectures, and report on the experiments we conducted with it to have a preliminary assesment of its performance in terms of execution costs and latency.

5.1   Prototype and Experimental Setting

We implemented our system based on the Ethereum blockchain. We encoded the on-chain components of our prototype in Solidity, the most used language for Ethereum smart contracts at present. We resorted to Node.js scripts to implement the off-chain components and the Web3 library to let them interact with the blockchain, i.e., for the subscription to event emissions and to send transactions to the blockchain. The produced code is openly available on GitHub.[4]

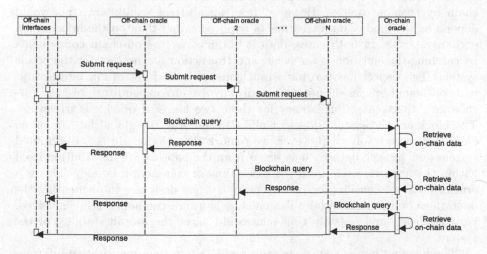

Fig. 6. Sequence diagram of the interaction with a decentralized pull-out oracle

---

[4] The implemented prototypes of the oracles used in the experiments are available at: https://github.com/DavideBasile1810355/Decentralized_Oracles/.

To run our tests, we deployed the on-chain components of our prototype on Ropsten,[5] an Ethereum public testnet, in order to execute the tests and obtain the needed information about latency and costs. The test phase took place through four different accounts used to deploy the smart contracts and send transactions from the off-chain components. The transactions involved in our experiments are identified by the interactions with the following contracts and can be retrieved via Etherscan: 0xd7c351Eb1DfaFCf19bf47D3fe55a9D761a274bd7; 0xA6a80830855c81b472A6aa9efb36bBA0fF36A5e4; 0x7Cc2d01fb411b9E59924f2Bc79002f93E9A44ddB; 0xAF69860c860A00d723fc0651f22637aF3b1B0d6D.

5.2   Performance Tests

Using our proof-of-concept implementation, we have conducted a preliminary assessment of its performance in terms of latency and costs, in an attempt to have a rough estimation of the differences between centralized and decentralized oracle architectures. A fully-fledged comparative study is out of scope for this paper and we envision it as a relevant aim for future studies.

The first important consideration that we made is about the outbound (pull and push) oracles. Although every on-chain computation requires the triggering of a transaction, we do not consider that transaction when measuring the performance of outbound oracles as they would not directly pertain to the oracle operations per se but rather to the pre-processing by the smart contract. Indeed, on-chain computation may be required to produce the data later retrieved by pull-out oracles, or for the production of the information to be transmitted off-chain by push-out oracles. However, from an abstract standpoint, this would depend on the kind of data treatments required rather than on the information exchange per se. In both cases, data is obtained by the off-chain components by catching the emission of an event, and this action has no cost for the oracle system. This aspect has two important consequences. First of all, interactions with outbound oracles do not necessarily involve any expenditure of gas. Furthermore, the transaction latency for these two kinds of oracles is irrelevant. The blockchain ecosystem does not affect in any way the global latency as no block mining is involved. However, we remark that blockchain is a distributed system and, as such, latency may occur from the information distribution itself within the network, aside from the block time or transaction latency. This is a crucial factor to consider for process-aware system designers implementing the operations on-chain: especially if numerous software components are involved, variable delays and possibly time-outs could affect the overall stability of the system.

The inbound implementations can provide interesting quantitative information that can be used for a preliminary performance assessment about costs and latency. We quantify the spending of the oracles in gas units and its equivalent amount in Euros. The exchange rate considered at the time of the experiments

[5] Rospten explorer: https://ropsten.etherscan.io/. Accessed: July 14, 2021.

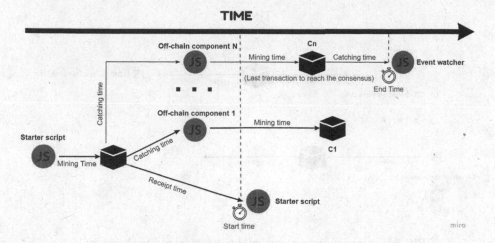

Fig. 7. Latency measurement for the pull-in oracle

is 495 Euros per Ether (ETH), while the average gas price considered for the ETH/gas conversion is 8 Gwei (0.000,000,008 ETH) per gas unit. Regarding the experiments on latency, our goal was to measure the time elapsed between the event that starts the interaction and the arrival at destination (i.e., the blockchain) of the information.

Figure 7 illustrates our time measurement scheme for the pull-in oracle. Considering our running example, the starter script represents the supplier's request to verify the creditworthiness of the customer. The pull-in oracle begins its execution as soon as the starter script receives the mining receipt of its request – then we start the timer. At that point, the off-chain component is activated by the on-chain component, and after it has retrieved the requested data from the off-chain environment, it sends a transaction to the on-chain oracle with that information. The end time of the measurement corresponds with the instant in which the on-chain component terminates the computation of the received input. In a decentralized scheme, oracles employ separate off-chain components that work independently. Therefore, the information processing from the on-chain component can begin only when the latest off-chain component has transacted its data.

Figure 8 depicts our measurement scheme for the decentralized push-in oracle (used in our running example for the delivery confirmation). The start time corresponds with the first transaction being sent by one of the off-chain components. The end time elapses when the latest confirmation receipt is received confirming the sending of aggregate data from the on-chain component. Notice that there is only one initial transaction and one final receipt in the centralized case.

Table 2 reports on the experimental results. For each test we executed 50 runs, totalling 200 runs (i.e., 100 for the centralized case and 100 for the decentralized case). Table 2(a) shows the results of the experiments for the cen-

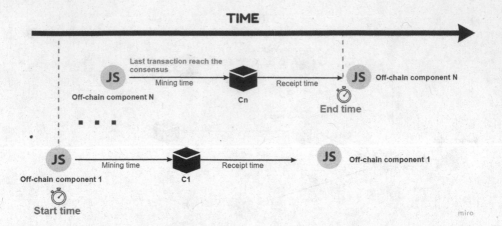

Fig. 8. Latency measurement for the decentralized push-in oracle

tralized implementations, reporting the mean, minimum and maximum values, and the standard deviation. As it turns out, the fastest implementation is that of the pull-in oracle, with a mean latency value of 18.24 s and a standard deviation of 15.23 s. The push-in oracle, instead, took 23.54 s on average with a standard deviation of 16.56 s. For as far as costs are concerned, the most expensive implementation is that of the pull-in oracle (with a mean of 42,000.98 gas units, while the push-in oracle required 39,505.47 units).

We evaluated the execution cost for oracles both in terms of the singular off-chain components (which we denote as Node 1, Node 2, and Node 3) and in terms of the whole oracle system. Considering the motivating scenario, both the pull-in and the push-in oracles employ three off-chain components that work independently. The cost of the single interaction is given by the sum of all the transaction costs, generated by each independent component (denoted as "C1", ..., "Cn" in Figs. 7 and 8). Table 2(b) shows the results for the decentralized case. As it can be seen in the table, in both cases independent nodes of the same system determine different mean costs. The test shows that some nodes require on average more gas than others although they belong to the same oracle. In other words, the gas consumption of the three off-chain components is not balanced. This can be explained by the order whereby the off-chain nodes send their transaction to the on-chain component. Indeed, the on-chain component provides the decentralized application with the data when all the off-chain components have sent their transaction. By considering the single run, the last off-chain node that sends the transaction containing the data is the one that spends more. In this case, the code executed by the transaction has a higher computational complexity because it also includes the operations for the delivery of the data to the smart contract of the DApp. The slowest decentralized implementation is the pull-in oracle a mean of 34.56 s, while the push-in oracle takes 27.78 s on average. Concerning the costs, the tests show that 179,175 units of gas are spent for the pull-in oracle and 115,667 units of gas for the push-in one.

Table 2. Latency and cost test results

(a) centralized oracles

	Mean	Min	Max	Std. dev.
Pull-in oracle				
Cost (gas)	42000.98	22550	62736	9039.54
Cost (euro)	0.17	0.09	0.25	/
Latency (seconds)	18.24	4.00	93.12	15.23
Push-in oracle				
Cost (gas)	39505.47	38003	42239	2027.04
Cost (euro)	0.16	0.15	0,17	/
Latency (seconds)	23.54	4.00	72.92	16.56

(b) Decentralized oracles

	Mean	Min	Max	Std. dev.
Pull-in oracle				
Cost (gas), Node 1	69232.97	25919	94816	24124.19
Cost (gas), Node 2	60565.05	22919	94794	27895.94
Cost (gas), Node 3	49377.23	25919	109522	27000.39
Total cost (gas)	179175.30	110300	236162	/
Total cost (euro)	0.72	0.44	0,95	/
Latency (seconds)	34.56	8.21	100.33	20.62
Push-in oracle				
Cost (gas), Node 1	42737.50	30098	58720	7808.17
Cost (gas), Node 2	42309	30098	58720	7846.29
Cost (gas), Node 3	30620.50	30098	58720	8034.68
Total Cost (gas)	115667	105338	136913	/
Total Cost (euro)	0.46	0.42	0.55	/
Latency (seconds)	27.78	5.79	69.65	14.34

6   Opportunities and Challenges

One of the main aspects of the proposed decentralization architecture is avail-ability. By decentralizing the structure of the oracle we eliminate a single point of failure. In the case of the centralization, if the unique control entity in charge of the information flow fails, the entire system oracle stops working and the commu-nication ends. On the contrary, in the proposed decentralized architecture, there is no central authority, since every off-chain component interacts independently with the on-chain component, which is, in turn, deployed on a decentralized system (the blockchain). In this way, the risk of failure for the whole system is reduced.

Another aspect that is affected by the decentralization is the integrity of data. Multi-party processes that rely on oracles may need to perform complex operations involving significant amounts of resources, and they cannot tolerate faulty or altered data. The fact that the information is not maintained by only one entity decreases the risk of counterfeit data injection (as in the unlocking of funds upon the confirmation from multiple nodes that the invoice was registered). In this way, reliability and trust could be generated. Of course, it is necessary to define internal mechanisms so that an agreement between the different com-ponents can be achieved. This specific aspect can increase the complexity of the whole oracle system, thus the centralized version might be preferred in some scenarios.

The centralized and the decentralized implementations allowed us to carry out a preliminary analysis of the performance (latency and costs) of the different kinds of architecture for the inbound oracles. As illustrated in Table 3, in all cases the decentralized prototypes require higher costs and cause more latency than their centralized version.

Table 3. Comparison table between centralized and decentralized implementations

	Centralized	Decentralized
Pull-in oracle		
Average cost (gas)	42,000.98	179,175.30
Average latency (seconds)	18.24	34.56
Push-in oracle		
Average cost (gas)	39,505.47	115,667
Average latency (seconds)	23.54	27.78

This can be explained by the presence of multiple transactions in the case of the decentralized versions. Indeed, our use case involves the definition of three off-chain components each of which generates one transaction for every procedure. Unlike the centralized versions that defines only one transaction for every information exchange. In this way, the mean cost of the system grows with the number of off-chain components involved. Alternatively, a decentralized system could check the agreement between the off-chain components in the real world, whereby only one transaction containing the final data would be generated. Therefore, it could serve as a viable alternative. However, if the entity in charge of sending the final transaction fails, the whole system stops working with such a solution, and the single point of failure problem persists.

Regarding latency, the difference between the two architectures is less evident. In the case of the pull-in oracle, the centralized version mean result is of 18.24 s against the 34.56 s of the decentralized version. The push-in centralized version, instead, generates a mean result of 23.54 s, against the 27.78 s of the decentralized version.

7   Conclusion and Future Work

In this paper, we investigated on the use and development of decentralized oracles as a means to enhance availability, integrity and trust of information exchanges between the blockchain and the outer environment in a business process context. We started with the design and development of on-chain components that communicate with the off-chain modules developed in a centralized version. Subsequently, we turned the oracles architecture into a decentralized one and compared it with the previous version. Our prototype was evaluated in terms of execution costs and latency.

In this paper, we have focused on the Ethereum blockchain in particular. Our study will be complemented with the development of oracles that are compatible with other blockchain platforms and then study the use of decentralized oracles for communication between multiple blockchains. Furthermore, the Solid Web[6] has been recently proposed as a paradigm for web applications preserving data

---

[6] Solid Web: https://solid.mit.edu. Accessed: July 14, 2021.

ownership and privacy. Reportedly, blockchain can be a key enabler of this novel paradigm [6]. Therefore, we will study the adoption of decentralized oracles to link decentralized systems and information producers and consumers to the Solid Web. Moreover, in this paper we have devised the merge and consistency-check of information exchanged with multiple off-chain components as an on-chain operation. Though more robust, this approach could incur higher costs than a purely off-chain mechanism. Therefore, an analysis of the best suitable trade-offs in terms of load-balancing and security of the two solutions is part of our envisioned future work. Finally, we will conduct in-depth studies on the scalability and robustness of the architecture, with an investigation on potential threats to security.

Acknowledgments. The authors are grateful to the reviewers for their precious feedback. The work of C. Di Ciccio was partially funded by the MIUR under grant "Dipartimenti di eccellenza 2018–2022" of the Department of Computer Science at Sapienza and by the Sapienza research project "SPECTRA".

References

1. Ahram, T., Sargolzaei, A., Sargolzaei, S., Daniels, J., Amaba, B.: Blockchain technology innovations. In: 2017 IEEE Technology Engineering Management Conference (TEMSCON), pp. 137–141 (2017)
2. Al-Breiki, H., Rehman, M.H.U., Salah, K., Svetinovic, D.: Trustworthy blockchain oracles: review, comparison, and open research challenges. IEEE Access **8**, 85675–85685 (2020)
3. Beniiche, A.: A study of blockchain oracles. CoRR abs/2004.07140 (2020)
4. Breidenbach, L., Cachin, C., et al.: Chainlink 2.0: next steps in the evolution of decentralized oracle networks (2021). https://research.chain.link/whitepaper-v2.pdf
5. Buterin, V., et al.: A next-generation smart contract and decentralized application platform. White paper 3(37) (2014)
6. Cai, T., Yang, Z., Chen, W., Zheng, Z., Yu, Y.: A blockchain-assisted trust access authentication system for solid. IEEE Access **8**, 71605–71616 (2020)
7. Dannen, C.: Introducing Ethereum and Solidity: Foundations of Cryptocurrency and Blockchain Programming for Beginners, 1st edn. Apress, New York (2017)
8. Diii Ciccio, C., et al.: Blockchain support for collaborative business processes. Inform. Spektrum **42**(3), 182–190 (2019)
9. Dumas, M., La Rosa, M., Mendling, J., Reijers, H.A.: Fundamentals of Business Process Management. 2nd edn. Springer, Heidelberg (2018). https://doi.org/10.1007/978-3-662-56509-4
10. Feig, E.: A framework for blockchain-based applications. CoRR abs/1803.00892 (2018)
11. Hens, P., Snoeck, M., De Backer, M., Poels, G.: Decentralized event-based orchestration. In: zur Muehlen, M., Su, J. (eds.) BPM 2010. LNBIP, vol. 66, pp. 695–706. Springer, Heidelberg (2011). https://doi.org/10.1007/978-3-642-20511-8_63
12. Hu, Y., et al.: The use of smart contracts and challenges. CoRR abs/1810.04699 (2018)

13. Ladleif, J., Weber, I., Weske, M.: External data monitoring using oracles in blockchain-based process execution. In: Asatiani, A., García, J.M., Helander, N., Jiménez-Ramírez, A., Koschmider, A., Mendling, J., Meroni, G., Reijers, H.A. (eds.) BPM 2020. LNBIP, vol. 393, pp. 67–81. Springer, Cham (2020). https://doi.org/10.1007/978-3-030-58779-6_5
14. Lo, S.K., Xu, X., Staples, M., Yao, L.: Reliability analysis for blockchain oracles. Comput. Electr. Eng. **83**, 106582 (2020)
15. Mammadzada, K., Iqbal, M., Milani, F., García-Bañuelos, L., Matulevičius, R.: Blockchain oracles: a framework for blockchain-based applications. In: Asatiani, A., et al. (eds.) BPM 2020. LNBIP, vol. 393, pp. 19–34. Springer, Cham (2020). https://doi.org/10.1007/978-3-030-58779-6_2
16. Mendling, J., Weber, I., van der Aalst, W.M.P., vom Brocke, J., Cabanillas, C., et al.: Blockchains for business process management - challenges and opportunities. ACM Trans. Manag. Inf. Syst. **9**(1), 4:1–4:16 (2018)
17. Moantly, D.: Ethereum for Architects and Developers: With Case Studies and Code Samples in Solidity. Apress, New York (2018)
18. Mühlberger, R., et al.: Foundational Oracle Patterns: Connecting Blockchain to the Off-Chain World. In: Asatiani, A., et al. (eds.) BPM 2020. LNBIP, vol. 393, pp. 35–51. Springer, Cham (2020). https://doi.org/10.1007/978-3-030-58779-6_3
19. Nakamoto, S.: Bitcoin: a peer-to-peer electronic cash system. Technical report (2008)
20. Nofer, M., Gomber, P., Hinz, O., Schiereck, D.: Blockchain. Bus. Inf. Syst. Eng. **59**(3), 183–187 (2017)
21. Nóbrega, T., Pires, C.E.S., Nascimento, D.C.: Blockchain-based privacy-preserving record linkage: enhancing data privacy in an untrusted environment. Inf. Syst. **102**, 101826 (2021)
22. Ohm, M., Plate, H., Sykosch, A., Meier, M.: Backstabber's knife collection: a review of open source software supply chain attacks. CoRR abs/2005.09535 (2020)
23. Tijan, E., Aksentijevic, S., Ivanić, K., Jardas, M.: Blockchain technology implementation in logistics. Sustainability **11**, 1185 (2019)
24. Xu, X., et al.: The blockchain as a software connector. In: WICSA, pp. 182–191. IEEE Computer Society (2016)
25. Xu, X., Pautasso, C., Zhu, L., Lu, Q., Weber, I.: A pattern collection for blockchain-based applications. In: EuroPLoP, pp. 3:1–3:20. ACM (2018)
26. Zheng, Z., Xie, S., Dai, H., Chen, X., Wang, H.: An overview of blockchain technology: architecture, consensus, and future trends. In: BigData Congress, pp. 557–564. IEEE Computer Society (2017)

Methods for Decentralized Identities: Evaluation and Insights

Walid Fdhila[1,2(✉)], Nicholas Stifter[1,2], Kristian Kostal[3], Cihan Saglam[4], and Markus Sabadello[4]

[1] Secure Business Austria (SBA-Research), Vienna, Austria
wfdhila@sba-research.org.com
[2] University of Vienna, Wien, Austria
[3] FIIT, Slovak University of Technology in Bratislava, Bratislava, Slovakia
[4] DanubeTech GmbH, Wien, Austria

Abstract. Recent technological shifts have pressured businesses to reshape the way they operate and transact. At the hart of this restructuring, identity management established itself as an essential building block in both B2C and B2B business models. Trustworthy identities may refer to customers, businesses, suppliers or assets, and enable trusted communications between different actors. Unfortunately, traditional identity management systems rely on centralized architectures and trust in third party services. With the inception of blockchain technology, new methods for managing identity emerged, which promise better decentralization and self-sovereignty. This paper provides an evaluation of a selection of distributed identity methods, and analyzes their properties based on the categorization specified in the W3C recommendation rubric.

Keywords: Blockchain · Distributed identity · Self-sovereign Identity · DID method

1 Introduction

In today's internet, organizations such as Google, Facebook or Amazon centrally manage and control vast amounts of cross-correlating data about individuals and their identities. An already diminished trust in such centralized systems by its users is further brought into question, as recent breaches have exposed their private data on a massive scale, urging the need for new *decentralized* methods that give individuals full control back over their data.

By providing the necessary infrastructure and renewed interest in Byzantine Fault Tolerance (BFT) [11], the advent of blockchain technology paved the way for such new decentralized methods for establishing trustworthy distributed identities that do not rely on a central entity serving as a single point of trust. This approach is called self-Sovereign Identity (SSI), in which entities or individuals become their own identity providers, thus creating and controlling one or multiple (i) decentralized identifiers (DIDs), and (ii) verifiable credentials (VCs)

© Springer Nature Switzerland AG 2021
J. González Enríquez et al. (Eds.): BPM 2021, LNBIP 428, pp. 119–135, 2021.
https://doi.org/10.1007/978-3-030-85867-4_9

[6]. (i) A DID is a unique identifier, usually associated with a DID document (also called continuation document) that specifies cryptographic material, verification methods and services essential for proving ownership of the DID and trustworthy communication with the DID owner. (ii) Verifiable credentials are identity attributes and assertions about a specific subject issued by an identity provider. In contrast to traditional credentials, a relying party (third party service) can check the validity of a VC without having to interact with the issuer. A DID method, on the other hand, defines how a DID can be created, resolved, updated and revoked. Currently, there exist over one hundred DID methods that rely on different architectural designs or infrastructures. Therefore, developing a use case that employs decentralized identifiers requires a good understanding of the properties of such methods and what they offer in terms of governance, security or operation. In this paper, we provide a qualitative evaluation of six DID methods following the guidelines of the W3C DID method rubric.[1]

The remainder of the paper is structured as follows. Section 2 introduces basic concepts, while Sect. 3 outlines the methodology. Section 4 evaluates the different DID methods, Sect. 5 discusses the results and concludes the paper.

2   Background

Decentralized Identifiers (DIDs). DIDs are unique identifiers whose purpose is to ensure trustworthy and persistent communication channels between entities. In contrast to common identifiers such as URIs, phone numbers or social media identifiers, which are issued and controlled by third parties, DIDs enable individuals and organizations to issue and control their own identifiers. DIDs, as specified in the W3C recommendation draft[2], are strings that have the following format: $did : <did\_method> : <method\_specific\_identifier>$, where did is the prefix, did_method refers to the method specification that defines the precise operations for creating, resolving, updating and revoking specific DIDs, and $method\_specific\_identifier$ is a unique identifier within the method. An example of a DID created using the $btcr$ method[3] on the Bitcoin blockchain would be: $did : btcr : 8kyt - fzzq - qpqq - ljsc - 5l$.

DID Document. Similar to DNS resolution where a URL is provided as an input and the corresponding IP address is returned as output, DID resolution takes the DID as input and returns a DID document as output. The latter contains, among others, cryptographic material (public keys for authentication, authorization, and interaction), verification methods for proving ownership of the DID, and service endpoints for enabling trusted communication with the subject, for instance to exchange verifiable credentials.

---

[1] This work expands upon a technical report that evaluates DID method specifications [4] which was conducted by the authors.

[2] See: https://www.w3.org/TR/did-core.

[3] See: https://w3c-ccg.github.io/didm-btcr/.

DID Method. The diversity in the blockchain ecosystem led to a plethora of different methods for creating and resolving DIDs. Although most of these methods comply with the W3C DID specification, each of them comes with specific properties and offers different guarantees depending on the underlying technology or governance framework. As such, each DID method specifies how corresponding DIDs are created, resolved, updated and deactivated (CRUD).

3   Methodology

3.1   DID Method Selection

At the time of writing, there exist over 100 DID methods that differ in various aspects such as (i) the underlying infrastructure (ii) governance, (iii) operation, and (iv) security. Unfortunately, a large number of these methods are either conceptual designs, unimplemented proposals or stale projects that are no longer maintained. Additionally, the focus is often placed on operational elements such as CRUD, while the remaining aspects of the method receive considerably less attention and documentation, rendering an evaluation difficult.

Based on the authors' expertise in the domain of SSI and their involvement in the implementation of the universal registrar and resolver for DIDs[4,5], a set of 6 DID methods, that cover different architectural designs, were selected and evaluated: (i) blockchain-based (ii) non blockchain-based, (iii) public permissioned, (iv) public permissionless and (v) pairwise DIDs. The selection, namely {DID:BTCR, DID:V1, DID:ETHR, DID:SOV, DID:WEB, DID:PEER} offers sufficient documentation and implementation details for a fair evaluation and covers approaches that are currently well received by the SSI community.

3.2   Evaluation Process

In this paper a qualitative evaluation is performed using the guidelines specified in the W3C DID method Rubric V1.0 (06 Jan. 2021)[6] and additional criteria derived from the principles of SSI [8]. The paper, therefore, provides both a comprehensive and comparative study of the DID methods and for each evaluation criteria considers three overlapping dimensions as follows:

(i) *Network* is the underlying communication layer, i.e., how and with whom users need to communicate to invoke the operations of the method.
(ii) *Registry* is a given instance of recorded state changes, managed according to the specification, using the communication layer.
(iii) *Specification* is the governing document of the method that defines and outlines how a particular method implements the required and any optional components of the DID core specification.

---

[4] See: https://dev.uniresolver.io/.
[5] See: https://github.com/decentralized-identity/universal-registrar/.
[6] See: https://w3c.github.io/did-rubric/.

Evaluation Criteria. The criteria are grouped into four categories, each focusing on a specific aspect of the method:

(1) *Rulemaking* captures the degree of decentralization in the governance of the DID method. It covers who can define the rules and how they are defined with respect to each of the aforementioned dimensions. For instance, the economic interest behind a DID method can impact its centralization if the goal is to support the interest of a certain group.

(2) *Operation* focuses on the CRUD operations and evaluates how the rules are executed. It also addresses the *openness* of the operation, i.e., whether it is restricted to a select group (permissioned) or open to participation by anyone (permissionless). Permissioned operation can impact the availability of the network to various participants, which affects inclusivity with regard to underserved or vulnerable populations. It may also expose the permission giver to legal ramifications.

(3) *Security* covers potential attack vectors against both, the integrity and correctness, as well as the privacy and self-sovereignty of users, that can arise through the method design choices and employed technologies.

(4) *Implementation* touches on aspects and challenges regarding an actual implementation and utilization of the corresponding DID method in practice.

The evaluation is conducted by five experts with diverse technical and theoretic backgrounds in distributed ledger technology in general and self-sovereign identity in particular. Some of the evaluators are also involved in standardization efforts by W3C for decentralized identifiers and verifiable credentials, in addition to projects for the implementation of a universal DiD registrar and resolver that supports around 50 DiD methods. As aforementioned, the selected DiD methods were chosen to cover different architectural designs and rely on various infrastructures that obey to different governance rules.

4   Evaluated DID Methods

4.1   did:btcr

Description. DID:btcr uses transactions on the Bitcoin blockchain for registering, updating and revoking identities. The DID corresponds to the transaction reference $TxRef^7$, which encodes details (i.e., chain, block height, transaction index and optionally outpoint index). The transaction can optionally include an OP_RETURN as part of the transaction outputs to refer to a DID document, otherwise, a default document is automatically created. OP_RETURN is a Bitcoin script opcode [2], which can be used to embed up to 80 bytes of data in a transaction. Updating the DID is achieved by spending the current outpoint and setting the OP_RETURN with a reference to the updated DID document. Reading a DID requires a lookup of the $TrxRef$ and following the chain of spending transactions until the last one with an unspent outpoint is reached. If the last transaction has no OP_RETURN, it means the DID has been revoked.

---

[7] See: https://en.bitcoin.it/wiki/BIP_0136.

Rule Making. *Network and Registry.* For did:btcr, the network and registry are actually the same and correspond to the Bitcoin protocol. Changes to the protocol require drafting a Bitcoin Improvement Proposal (BIP) which is then openly discussed within the community. Therefore, participation in network governance is open and anyone can join, comment and contribute to open debate (open contribution). The process towards BIP acceptance follows the guidelines defined in BIP:2, where it is recommended that the acceptance of a BIP requires at least a 95% acceptance rate by the miners of the last 2016 blocks, unless there is rationale. The deciding group is not closed and includes known and unknown entities/miners (breadth of authority). However, participation in governance cannot be considered fair, as miners with higher hash power have more influence in the decision making. In DID:btcr, although miners receive block rewards and transaction fees, the governance of the DID method itself is decentralized and therefore, established to the public good.

Specification. The specification of the DID:btcr is created and maintained by a closed set of contributors. Comments and suggestions to the specification are open, but the decision lies within a closed and known set of people. Although participation in the specification governance requires time and effort, no incentives to the specification governing entity are defined, thus confirming prior conclusions on financial goals of the method, which is established for the public good without extracting rents or remunerations.

Operation. As the Bitcoin blockchain is public and permissionless, where anyone can participate, read and write (transact) with the ledger, the operations of the did:btcr also do not require any permission. However, resolving a DID without relying on authoritative intermediaries requires operating a full node, which can prevent the use of resource constrained edge devices to directly resolve DIDs. For registering DIDs edge devices can be sufficient unless a continuation document is specified, which would require additional resources for hosting it (e.g., a server). Note that did:btcr supports the creation of both universal and paired DIDs. Although Bitcoin is public and, in principle, anyone can resolve a DID, it is also possible to create a default DID (without setting the OP_RETURN), which would make it indistinguishable from a normal transactions, and therefore can be used in a pairwise manner. It is possible for anyone to retrieve a cryptographic proof of the history of changes (transactions), thus theoretically enabling public auditability. However, referencing continuation documents that reside on mutable storage can hinder these benefits.

Security. While relying on Bitcoin transactions as DIDs ensures integrity and persistence, referring to continuation documents raises concerns over censorship and mutability. It is possible for both the storage provider or a governing entity to censor the access to the server hosting the continuation document, rendering the resolution of the corresponding DID impossible without first updating it. Besides censorship, failure of the host server directly impacts the availability of the DID (for resolution) unless a caching mechanism is implemented. Integrity of a continuation document can be checked as it is signed with the transaction input key. However, a discrepancy between the specification document and the design decision document currendly renders it unclear which input key to use for signing the continuation documents, i.e., (i) the DID creation input key, or (ii) keys used for updates. The latter case would open up the possibility of claiming someone else's DID under certain conditions [4].

Implementation. One challenge with implementing btcr DIDs is that there is no single definitive, complete, and up-to-date specification. Implementers have to combine information from different sources, ask questions from the community, and analyze existing examples and code to build a compatible implementation. CRUD operations on btcr DIDs can be difficult when a continuation DID document is needed, since this requires access to a web server, in addition to access to the Bitcoin blockchain. For resolving btcr DIDs, the main challenge is "following the tip", i.e., the process of looking up unspent outpoints of a transaction after a btcr DID has been updated or deactivated, which is not readily supported by all Bitcoin implementations. Finally, an aspect of btcr DIDs is that during creation, the actual DID only becomes fully known and stable after a transaction is mined in a block and sufficiently confirmed. This requires implementations to maintain some kind of internal state and monitoring process.

4.2 did:sov

Description. Sovrin is a private non-profit foundation and its DID method relies on a public permissioned blockchain specifically and exclusively targeted for self-sovereign identity [10]. Sovrin's technical underpinnings derive from Hyperledger's Indy project and it employs the plenum protocol, which is an enhancement of RBFT (Redundant Byzantine Fault Tolerance) [1].

Rule Making. *Network and Registry.* In did:sov, the network and the registry both correspond to the Sovrin network. Governance is restricted to a board of trustees (BoT) that decide on (i) how the network evolves, and (ii) approval of new stewards or governance proposals by the Sovrin governance framework working group (SGFWG). Stewards are independent entities (e.g., universities, organizations) responsible for endorsing transactions and writing to the ledger, and have to comply with the governance framework approved by the BoT. All governance documents are open to public review and comment, but the decision is ultimately restricted to a closed group (i.e., the BoT). Similar to DNS governance, despite being a non-profit public organization, the Sovrin foundation collects fees and rents to ensure economic viability of the infrastructure. Additionally, although there exist several and different governance bodies within the Sovrin foundation (e.g., SGFWG, STGB, EAC) and any one can join, the ultimate governance approval remains under the BoT control. To be part of the BoT, one should first be nominated by the nomination/transition committee and then voted by the current BoT. Participating in governance is clearly restricted and requires modest costs in terms of efforts and time.

Specification. The specification is governed by the Sovrin technical governance board through the Sovrin trust framework, and revisions (called controlled documents) should undergo the BoT approval.

Operation. In contrast to did:btcr, while anyone can read from the Sovrin ledger, writing to it is permissioned and restricted to transaction endorsers (e.g., stewards). Sovrin publicly shares their annual financial reports, which can be checked by anyone, thus rendering its financial accountability transparent. To resolve DIDs, did:sov does not require implementing a full node, but can instead rely on state proofs making the use of edge devices with limited resources possible. However, registering a DID without relying on intermediaries is not possible as it has to be achieved through an endorser. Finally, the public nature of Sovrin enables anyone to retrieve a cryptographic proof of all changes to a DID document, making the system auditable.

Security. In practice, did:sov is censorship resistant as it relies on a distributed network of nodes (stewards) from all over the world, responsible for accessing and writing to the ledger. As such, a user that is censored by a steward can readily register a DID using a different one. However, this does not prevent the BoT from issuing new endorsement policies that censor specific type of users, with which endorsers have to comply. The integrity of the ledger is maintained by the diverse stewards and observer nodes, and can also be publicly verified using for example anchored state proofs. Although confidentiality is not a required property, it is possible in did:sov to create pairwise DIDs that are not stored on the ledger. Besides, Sovrin also supports the creation of blinded DIDs using zero knowledge proofs. According to the Sovrin GDPR compliance policies, personal data may not be written to the ledger (data minimization), i.e., only public DIDs and the corresponding DID documents, credential definitions and revocation registries are stored on the ledger. Sovrin also uses software agents (e.g., edge agent, cloud agent) to store and manage credentials and keys, and communicate with other agents in a peer to peer fashion using the didcomm protocol.

Implementation. This method was designed for Sovrin, which is widely known and has been used by many implementers and real-life projects. It has also implicitly been designed for other ledger instances of Hyperledger Indy. In practice however, applications and services building on top of these ledgers use custom identifier and discovery formats instead of actual DIDs and DID documents, which has made this DID method hard to understand. A DID method specification exists, but it is outdated. Implementations of sov DIDs therefore are currently mostly based on community knowledge and undocumented assumptions. The ledger itself offers basic operations such as NYM and ATTRIB that make it possible to build DIDs and DID documents on top. The shortcomings and confusion around implementing sov DIDs is expected to be solved with the arrival of the new Indy DID method.

4.3   did:ethr

Description. The did:ethr method is similar to did:btcr in that it also builds upon on a blockchain technology, in this case Ethereum. However, while Bitcoin employs a UTXO-based ledger design, Ethereum utilizes an account-based model and supports quasi-Turing complete *smart contracts* [7]. did:ethr leverages these properties by mapping Ethereum (externally owned) account addresses, which are derived from the public key of an ECDSA Secp256k1 asymmetric key pair, to identities. An important design decision of did:ethr is that the creation of a DID does not necessitate submitting a transaction to the Ethereum network. Instead, any regularly generated externally owned account address is considered a DID. In addition, Ethereum's smart contract functionality is used to realize a registry for CRUD operations and to allow for the delegation of control over an identity. Hereby, the smart-contract-based registry follows the preliminary ERC-1056 standard defined in Ethereum improvement proposal EIP-1056[8] and inherits desirable properties such as immutability, trustless execution and decentralization, from the underlying platform. While did:ethr appears to offer a lightweight and cost effective method for creating DIDs, the design choices introduce some unclear properties regarding the creation and revocation of DIDs that are never committed to through a transaction on the blockchain.

Rule Making. *Network and Registry.* Similarly to did:btcr, the network and registry in this method are also provided through the underlying blockchain system. However, did:ethr relies on a registry that is governed by a smart contract which can publicly be interacted with through any Ethereum account or other smart contract code. In regard to rule making, the current smart-contract-based registry specification renders the functionality *immutable* as by the properties of the underlying ledger. However, it is unclear if the current draft ERC-1056 registry design will change to include some ability for governance once the draft is finalized. For the basic network (blockchain), while anyone can participate fully in principle, to be able to meaningfully partake in Ethereum's consensus protocol requires significant hashrate and therefore financial resources. In practice high-hashrate proof of work blockchains such as Bitcoin and Ethereum present themselves more like networks where consensus is permissioned, as the average user has no realistic chance of influencing consensus decisions. In Ethereum, the governance authority is an open set of multiple parties and the process, in analogy to Bitcoin's BIPs, is governed through Ethereum Improvement Proposals (EIPs). The operational costs of the registry are fully transparent because they

---

[8] https://eips.ethereum.org/EIPS/eip-1056.

are publicly visible on the blockchain, and the network costs such as mining rewards and hashrate can also be deduced from on-chain data.

Specification. The did:ethr specification was initially created by uPort [9], however it is not use-case specific or geared toward an extraction of rent and established as a public good. It is openly available in GitHub and is currently governed by the Decentralized Identity Foundation, however there is no reason to assume that this excludes others from actively participating. The smart contract code of the registry is also publicly available and can be checked against the deployed contract.

Operation. The creation of ethr DIDs does not require any permission, special hardware requirements or even access to the full blockchain ledger, in particular if no transaction to the registry (e.g. for delegating control) is required. For some of the CRUD it can be necessary to interact with the on-chain smart contract through transactions that need to pay transaction fees to miners. These transaction costs are public and recorded in the blockchain. Overall the compensation to miners in Ethereum is highly transparent as all on-chain flows of cryptocurrency funds are publicly accessible. Reading the registry is possible for anyone (either through a light or full node or a third party service). The scheme is designed to be operational on different EVM compatible networks (e.g. Ethereum testnets, Rootstock etc.) and also allows to specify alternative registry addresses. Analogous to did:btcr, if all of the CRUD operations of the DID are performed through transactions, it is possible for anyone to retrieve cryptographic proof of these changes, enabling public auditability. On the other hand, if DIDs are not added to the registry through transactions, they can be used in a pairwise manner and also may offer some degree of privacy e.g., against metadata collection.

Security. The advantage of did:ethr lies in the design of not having to commit to the DID in a transaction unless the owner desires so or wants to include properties such as delegating control of the DID to another address. This however means that a DID that has been deleted/revoked which has never used the registry can not be distinguished from one that is not revoked. Integrity is ensured through blockchain consensus and the use of established algorithms for asymmetric cryptography and hash functions. Users can create their DID trustlessly by generating a Secp256k1 keypair. Utilization of the registry is trustless as long as the underlying ledger remains "permissionless", i.e. transactions are not censored by miners and are economically viable. Updates to the DID in the registry are publicly visible and in principle personally identifying information could also be encoded in the DID entry, introducing potential privacy issues. DID resolution and reading the registry can be done with good confidentiality as blockchain state is publicly and anonymously accessible, either through running one's own full or light Ethereum node or, with more trust assumptions, through the API of a third party provider. Similarly to the continuation documents in did:btcr, the method allows for "service endpoints" which can reference external, mutable resources such as URLs, thereby opening up potential concerns over censorship, persistence and privacy.

Implementation. This method can render it cost effective for anyone to create DIDs, as the creation step only requires the generation of a cryptographic key pair, without having to perform blockchain transactions. Resolving ethr DIDs requires read operations against Ethereum or the respective Network in which the registry smart contract is located, which can however readily be achieved with any standard Ethereum tools and only incurs modest resource requirements. Once update and deactivate operations are necessitated, implementers need to be able to write to the blockchain, which requires appropriate infrastructure to be in place (similar to other blockchain-based DID methods) and is subject to transaction fees.

4.4   did:web

did:web is a method that uses domain names as identifiers. The DID is a URI that points to a DID document stored on a web host server and registered within a DNS registrar. To resolve the DID, a HTTP GET request on the HTTPS URL generated from the DID is required. Updating the DID is achieved by replacing/updating the DID document on the hosting location. Revocation occurs if the DID document is deleted from the web host.

Rule Making. *Network and Registry.* Because the did:web method uses domain names to represent DIDs, both the DID network and registry correspond to the registries and registrars running the domain name servers. Therefore, evaluating governance aspects requires a thorough understanding on how such a traditional DNS system is governed. Governance of DNS is mainly the responsibility of the ICANN, a multistakeholder, private, non-profit organization that follows a bottom-up, consensus driven, model to coordinate the assignment of internet domain names and IP addresses [5]. Although each DNS registry/registrar might have separate internal rules and is responsible for allocating/selling its corresponding domain names, ultimately they have to comply and fulfill the agreements and policies of ICANN. The DNS model combines both public and private economies. The ICANN itself is a non-profit organization that acts for the common good of the public, but extracts registration fees from registries, registrars and indirectly registrants, for covering the running costs of the organization. However, from the perspective of registry and registrars, they extract rents to enhance their profits.

Specification. The specification is published by the credentials community group. Anyone can comment and raise issues through the specification GitHub repository, however, decision making is not clear and seems to be conducted by the specification authors.

Operation. While anyone can read, writing to the registry/DNS server is permissioned. Creating a DID requires a subscription within a registrar/registry or a third party seller, in addition to a web host for storing the did document (except the case where users host their own web servers). As such, to resolve a DID using the did:web method without relying on intermediaries, a user has to be an accredited registry. Besides the fact that this does not seem as a practical solution, it will also require exceptional resources and has to follow and comply with complex procedures. Furthermore, while domain names are meant to be used universally (unless using local/private network and DNS server), DID documents are stored on web hosts with no means of cryptographically proving the history of their changes, thus rendering auditability almost impossible.

Security. In did:web, censorship can happen at different levels: (i) the DNS or (ii) the web host. While a migration to a new web host would solve the latter scenario (against migration and update fees), the registry has still full control on removing or censoring specific DIDs. Although it is also possible to transfer the DID to another registrar, the registry (e.g., Verisign for .com) still has the power to deny the user request. In the current specification of did:web, it is not clear how integrity is addressed although a proposal to use hashlinks is suggested. Confidentiality, on the other hand, depends on whether or not the registration within the registrar is private. If not, the user has to reveal her basic information, thus giving registrars the ability to correlate the actual identity with the corresponding DID.

Implementation. Resolving web DIDs only requires a simple HTTP GET operation, which can be readily achieved in any programming language or operating system. Creating, updating, and deactivating only require storing and updating a file on a web server.

4.5 did:v1

Description. Veres One[9] is a project specifically targeted at the creation and management of DIDs. The v1 specification was drafted by members of Digital Bazaar and is hosted/maintained by the W3C Credentials Community Group. It relies on custom distributed ledger technology, which is based on the "Continuity"[10] BFT consensus protocol that appears specifically targeted for an application in Veres One. The Method is designed to extract rents and remuneration for its operators and it specifies a detailed governance structure for defining the relevant operational entities, governing bodies and how the method specification may be updated. At the time of writing, the collection of specification and design details regarding did:v1 presented itself challenging, as the documentation and code is spread over multiple GitHub repositories and websites and does not paint

---

[9] Cf. https://veres.one/.
[10] Cf. https://github.com/digitalbazaar/bedrock-ledger-consensus-continuity.

a coherent picture. Further, while the project management, governance and its goals are outlined, the presented structure is relatively complex and it is unclear how to readily verify if the project adheres to the specification and its claimed goals in practice.

Rule Making. *Network and Registry.* v1 intends to use a public ledger where, in principle, anyone can create and resolve DIDs. To keep in line with GDPR compliance, some elements of the DID can exist off chain. While it is claimed that the network is permissionless, operational details suggest that this property may only extend toward the ability of reading the ledger. Specifically, the employed novel consensus protocol and its ability to support an open participation model has not yet received sufficient peer-review to allow for an objective evaluation. The cost for participating as a network node in the Veres One network is not fully clear, but there will be at least a modest cost involved.

Specification. It appears that interested parties can contribute and participate, either by taking on a governance role, or commenting on the public GitHub repositories. However, ultimately control over the specification is held by the Veres One governing body and the entities controlling these repositories.

Operation. The software necessary to run a client is open source, requires minimal resources and can query information from network nodes. For trustless DID resolution users would have to run a network node themselves. In regard to creating or updating DIDs (writing to the ledger) did:v1 follows a model where users pay an *accelerator fee* that is distributed among the maintainers and participants of the network. Hence, did:v1 offers a more restricted permissioned model similar to did:sov. However, it appears to be possible to circumvent accelerators by performing a proof-of-work or partaking in the protocol as a consensus node.

Security. Within did:v1 it is currently unclear if the method can achieve its stated properties in a fully "permissionless" setting in practice, as the consensus protocol is not yet sufficiently analyzed. On the one hand, under the assumption that the ledger and its consensus protocol is fully permissionless, it can achieve censorship resistance. Public verifiability is also possible, however the method also supports external resources which may not be verifiable or could be censored. On the other hand, if consensus is only achieved by assuming a restricted set of participants, i.e., it is permissioned, it opens up the possibility of censorship. According to the method specification, GDPR compliance is achieved but it is not fully clear how this property is enforced in practice. More specifically, to be fully GDPR compliant consensus nodes need to verify that no personally identifiable data has been encoded in the DID, be it intentional or unintentionally. In relation to the right to be forgotten it may be necessary to delete entries in the blockchain's history. However, secure redactable blockchains in the permissionless setting are still a subject of ongoing research [3].

Implementation. The did:v1 method directly builds on JSON-LD and Linked Data Proofs, which can provide some familiarity for implementers. Resolving v1 DIDs is straightforward, since each network node exposes an HTTP GET interface for retrieving a fully compliant DID document. One primary question that remains open is the future evolution and implementation of the Veres One ledger. This includes both, whether the envisioned technological goals that are laid out in the specification can be achieved in practice, as well as how governance and network participation (e.g., permissionless or permissioned) is then realized depending on these technologies.

4.6   did:peer

Description. The core concept behind the did:peer specification hinges on the insight that there exist two categories of DIDs, namely *anywise DIDs* and *N-wise DIDs*. The former are intended to be used with an unknown number of parties whereas the latter are only intended to be known by exactly N enumerated participants, and the did:peer method addresses this type of DID. A *pairwise DID* is the special case where $N = 2$. N-wise DIDs are only relevant to its corresponding members and aspects such as resolution should only concern the involved parties. Hence the bulk of interactions can be moved off-chain with the possibility of connecting back to a chain-based ecosystem if needed.

Rule Making. *Network and Registry.* There are no specific networks for rule making, communications can go through any network channel. The decision on picking a specific network is up to the involved parties participating in did:peer and the registry is only at the peers, held locally. If the peers decide to change network picking and rules or registry, it is up to them. The creator of the DID is only one peer or a pair of peers which agrees on some rules, everything is peer-related. As the network and registry are created between a set of peers, it is only for the common good of those participants.

Specification. The did:peer specification is openly available on Github where anyone can propose improvements or changes, however the board of contributors who can accept such changes seems to be a closed group. It does not appear that the specification is geared toward the extraction of rent and is for the public good.

Operation. Anyone can, in principle, participate if she is a peer. However, the network or communication layer is visible only to peers participating in the operations if not otherwise decided by the involved peers. The registry is established between communicating peers and held locally, requiring little overhead or unnecessary data. The network and registry can be anything on what peers agree upon. and DIDs can be created and used contextually, between any set of parties. Auditability of operations depends on the concrete capabilities of the underlying registry and network that was agreed upon.

Security. The corresponding DIDs in did:peer are generated in a securely random process. This prevents attackers from discovering patterns in peer DIDs that might undermine privacy. Normally, peer DIDs are not persisted in any central system, so there is no trove to protect. However, in communication with dynamic peers, there is a special layered mechanism which is used to persist others' peer DID docs into backing storage which can be a ledger. Messages in this protocol are sent encrypted, by the specified format DIDComm's encryption envelope. This gives strong guarantees about the confidentiality and integrity of exchanged data. As the communication is mainly between two peers the needed security measures are partially minimized from the network point of view.

Implementation. Implementing peer DIDs takes some significant effort for implementers, since this DID method introduces a lot of new concepts that many developers will not be familiar with. On the other hand, the DID method renders it possible to progressively implement more features. Creation of a peer DID only requires generating a key pair, while other operations work via a peer-to-peer protocol between agents. It is not completely clear how peer DIDs currently fit in with other community developments, such as Hyperledger Aries and DIDComm.

5   Discussion and Conclusion

Blockchain technologies have opened up manifold opportunities toward realizing SSI systems that do not need to rely on centralized entities. While, in part, this is achieved through the intrinsic properties of these technologies, e.g., immutability, resistance to censorship, and decentralization, the degree to which a DID method can be considered decentralized or secure also depends on many other aspects and design choices. One cannot assume that just because a DID method is based on blockchain technology implicitly renders it decentralized. Indeed, it is extremely important to consider all dimensions, i.e., (i) network, (ii) registry, and (iii) specification, and assess a method's fundamental properties (e.g., governance, economic model or security) against each of these dimensions. Table 1 provides a comparative overview of the investigated methods. While *protection* determines a method' resistance to censorship, *persistence* evaluates the longevity of decentralized identifiers. *Integrity* ensures that DiDs and the corresponding DiD documents have not been tampered with, and *confidentiality* means that the DiD method gives the option to protect DiDs or DiD Documents from unauthorized disclosure if required. Finally, *Decentralization* examines how decentralized is a DiD method by evaluating the decentralization of its underlying network, registry and specification governance, and operations. For example, our evaluation has shown that despite did:btcr relying on the decentralized Bitcoin network for creating DIDs, the corresponding DID documents are still hosted on mutable storage, thus hindering blockchain benefits and introducing new risks of censorship, availability and persistence. Similarly, while did:ethr also builds upon a public permissionless Blockchain (Ethereum), its design relies on an on-chain ERC-1056 smart contract to manage and govern the DID registry.

By doing so, trust is shifted to both, the smart contract implementation as well as the underlying Ethereum blockchain. Note, that in both methods changes to the specification cannot prevent that operations on DIDs can also follow previous specification versions. Resolving them would hence require DID resolvers to maintain all previous resolution implementation versions. It is noteworthy to point out that most of the evaluated methods have a distinct lack of version control/migration mechanisms to prevent old DIDs from becoming unresolvable or obsolete. A clear definition on how to upgrade the method specification and assessment of its impact on the current implementation would clearly prove beneficial for introducing new features and mitigating security or performance issues.

There exist security, usability and scalability trade-offs between methods employing identity-specific ledgers and methods relying on public blockchains, and between the permissioned and permissionless operation of the underlying ledger. Identity-specific permissioned ledgers that rely on BFT-based consensus mechanism (e.g., Sovrin and Veres One) offer the advantage of better scalability and performance over traditional Blockchain designs (e.g., Bitcoin and Ethereum), however this comes at the cost of reduced decentralization and an increased risk of censorship by operators. This derives from the fact that the entities responsible for writing to the ledger have to comply with endorsement agreements that, in the end, might be changed by the governing entity, which to a certain extent is less decentralized and may serve specific interests.

Table 1. Comparative table of DID methods

	RuleMaking			Operation		Security*			
	Network	Registry	Specification	Network	Registry	Pro	Per	Int	Conf
did:btcr	● □	● □	○	● □	● □	+	+	+	±
did:v1	◑†	◑†	○	◑†	◑†	−	±	±	+†
did:ethr	● □	N/A†	◑	● □	● ■	+	+	+	−
did:sov	◑ ■	◑ ■	◑	◑ ■	◑ ■	−	±	+	±
did:web	◑	◑	○	◑ □	◑ □	−	−	−	±
did:peer	●	●	○	◑†	◑†	±	−	−	+

*Security - Pro: protection   Per: persistence   Int: integrity   Conf: confidentiality
● fully decentralized   ◑ partially decentralized   ○ centralized   N/A† not applicable
Required resources: ■ modest   □ substantial
†Not clear or well defined how method satisfies criteria at time of writing

The selected methods interestingly rely on different economic models that range from non-profit organizations which extract rents from the DID methods to totally open and free (not considering network fees) community projects. This creates a trade-off between the sustainability and growth of the project and trust in the system. Indeed, participation in governance and maintenance often requires substantial efforts and time. Hence, a failure to consider aligning incentives or covering operational costs in the method's economic model may lead to a stale project or to an outdated specification. On the other hand, incentives should not be aligned to only serve the economic interest of a specific entity, thus diminishing trust, openness, transparency and accountability of the system.

We hereby point out some of the challenges encountered while conducting this evaluation. First, the amount and quality of available documentation, as well as the discrepancy between some of the methods' specifications and their actual corresponding implementations, introduced uncertainties on how to fairly evaluate specific properties of the method. Moreover, some of the specifications might have changed during and after the evaluation, thus requiring continuous revision of the evaluation. Finally, some of the constructs and goals of specific methods are difficult to verify in practice or have yet to be implemented, leaving an answer to whether or not they can achieve the promised guarantees unclear.

To conclude, there is no clear winner among the evaluated DID methods. Each comes with advantages and disadvantages, and the selection of a particular method heavily depends upon the use case (e.g., supply chain, KYC, automotive process) and desired properties. Some application areas may require scalable and private systems, while others can necessitate a focus on distribution and trust. Furthermore, while blockchain offers unique security and decentralization properties for DID methods, it does not prevent flawed specifications and governance designs from introducing vulnerabilities that could jeopardize potential benefits.

Acknowledgments. This research is based upon work partially supported by (1) the Christian-Doppler-Laboratory for Security and Quality Improvement in the Production System Lifecycle; The financial support by the Austrian Federal Ministry for Digital and Economic Affairs, the Nation Foundation for Research, Technology and Development and University of Vienna, Faculty of Computer Science, Security & Privacy Group is gratefully acknowledged; (2) SBA Research (SBA-K1); SBA Research is a COMET Center within the COMET – Competence Centers for Excellent Technologies Programme and funded by BMK, BMDW, and the federal state of Vienna. The COMET Programme is managed by FFG. (3) the FFG ICT of the Future project 874019 dIdentity & dApps. (4) the FFG Industrial PhD project 878835 SmartDLP. (5) the U.S Department of Homeland Security Science and Technology Directorate's Silicon Valley Innovation Program (SVIP) under OTA 70RSAT20T00000030. Any opinions contained herein are those of the author(s) and do not necessarily reflect those of DHS S&T

References

1. Aublin, P.L., Mokhtar, S.B., Quéma, V.: RBFT: redundant byzantine fault tolerance. In: 2013 IEEE 33rd International Conference on Distributed Computing Systems, pp. 297–306. IEEE (2013)
2. Bartoletti, M., Pompianu, L.: An analysis of bitcoin op_return metadata. In: International Conference on Financial Cryptography and Data Security, pp. 218–230 (2017)
3. Deuber, D., Magri, B., Thyagarajan, S.A.K.: Redactable blockchain in the permissionless setting. In: 2019 IEEE Symposium on Security and Privacy (SP), pp. 124–138. IEEE (2019)

4. Fdhila, W., Stifter, N., Kostal, K., Saglam, C., Sabadello, M.: DID methods evaluation report - draft, January 2021. https://docs.google.com/document/d/1jP-76ul0FZ3H8dChqT2hMtlzvL6B3famQbseZQ0AGS8/

5. Klein, H.: ICANN and internet governance: leveraging technical coordination to realize global public policy. Inf. Soc. **18**(3), 193–207 (2002)

6. Lesavre, L., Varin, P., Mell, P., Davidson, M., Shook, J.: A taxonomic approach to understanding emerging blockchain identity management systems. CoRR abs/1807.06346 (2019)

7. Luu, L., Chu, D.H., Olickel, H., Saxena, P., Hobor, A.: Making smart contracts smarter. In: Proceedings of the 2016 ACM SIGSAC Conference on Computer and Communications Security, pp. 254–269 (2016)

8. Mühle, A., Grüner, A., Gayvoronskaya, T., Meinel, C.: A survey on essential components of a self-sovereign identity. CoRR abs/1807.06346 (2018)

9. Naik, N., Jenkins, P.: uPort open-source identity management system: An assessment of self-sovereign identity and user-centric data platform built on blockchain. In: 2020 IEEE International Symposium on Systems Engineering, pp. 1–7 (2020)

10. Reed, D., Law, J., Hardman, D.: The technical foundations of Sovrin. Technical report, Sovrin, 2016 (2016)

11. Stifter, N., Judmayer, A., Weippl, E.: Revisiting practical byzantine fault tolerance through blockchain technologies. In: Security and Quality in Cyber-Physical Systems Engineering, pp. 471–495. Springer, Cham (2019). https://doi.org/10.1007/978-3-030-25312-7_17

Author Index

Printed in the United States
by Baker & Taylor Publisher Services